AP* Physics 2 Essentials

An APlusPhysics Guide

Dan Fullerton

Physics Teacher
Irondequoit High School

Adjunct Professor
Microelectronic Engineering
Rochester Institute of Technology

Dedication

To my parents, for supporting all my crazy meanderings.
To those who recognize my errors, and allow me to fix them.
And of course to my girls, without whom my knowledge of
princesses, fairies, and unicorns would be severely lacking.

Credits

Thanks To:
Bob Enck, Paul Sedita, Christopher Becke, Peter Geschke, Mike Powlin, Laurie Peslak,
Rob Spencer, Doni Parnell, Dan Burns, Gardner Friedlander, Dolores Gende, James Phillips,
Brian Heglund, Joe Kunz, Karen Finter, Monica Owens, Jen Brooker, Joshua Buchman

Silly Beagle Productions
Webster, NY
Internet: www.SillyBeagle.com
E-Mail: info@SillyBeagle.com

Cover Design:
Interior Illustrations by Dan Fullerton, Jupiterimages and NASA unless otherwise noted
All images and illustrations ©2015 Jupiterimages Corporation and Dan Fullerton
Edited by Joe Kunz

*AP, AP Physics 2, Advanced Placement Program, and College Board are registered trademarks of
the College Entrance Examination Board, which was not involved in the production of, and does
not endorse, this product.*

Sales and Ordering Information
http://www.aplusphysics.com/ap2
Sales@SillyBeagle.com
Volume discounts available.
E-book editions available.

Printed in the United States of America
ISBN: 978-0-9907243-1-5

1 2 3 4 5 6 7 8 9 0 9 8 7 6 5 4

Silly Beagle Productions

Welcome to <u>AP Physics 2 Essentials - an APlusPhysics Guide</u>. From mechanics to momentum and sound to circuits, this book is your essential physics companion to complement AP Physics 2 lectures and physics textbooks, or it can be used as a high-level review book for your entire AP Physics 2 course. What sets this book apart from other review books?

1. It reviews the fundamental concepts and essential understandings required for success in AP Physics 2, and is written specifically to assist students in their AP Physics 2 courses and on the AP Physics 2 exam.
2. It focuses on basic concepts and relationships, items that can be taught reasonably well with a book, leaving development of deeper understandings and problem-solving skills where they belong, in the active classroom.
3. It includes more than 400 sample questions with full solutions, integrated into the chapters immediately following the material being covered, so you can test your understanding. You'll also find more than 60 AP-style problems at the end of the book.
4. It is supplemented by the free APlusPhysics.com website, which includes:
   a. Videos and tutorials on key physics concepts
   b. Interactive practice quizzes
   c. Discussion and homework help forums supported by the author and fellow readers
   d. Student blogs to share challenges, successes, hints and tricks
   e. Projects and activities designed to improve your understanding of essential physics concepts in a fun and engaging manner
   f. Latest and greatest physics news

Just remember, physics is fun! It's an exciting course, and with a little preparation and this book, you can transform your quest for essential physics comprehension from a stressful chore into an enjoyable and, yes, FUN, opportunity for success.

How to Use This Book

This book is arranged by topic, with sample problems and solutions integrated right in the text. Actively explore each chapter. Cover up the in-text solutions with an index card, get out a pencil, and try to solve the sample problems yourself as you go, before looking at the answer. If you're stuck, don't stress. Post your problem on the APlusPhysics website (http://aplusphysics.com) and get help from other students, teachers, and subject matter experts (including the author of this book!). Once you feel confident with the subject matter, test yourself and see how you perform. Review areas of difficulty, then try again and watch your understanding improve!

* There are topics presented in this book which are typically included in an introductory physics course but are not specifically tested on the AP Physics 2 Exam. These topics are marked with an asterisk to allow those focused solely on preparing for the AP Physics 2 Exam to bypass them in favor of topics that they know will be tested. Occasionally topics in this book are discussed slightly beyond the depth of content knowledge required for the AP Physics 2 Exam. This is done in an effort to provide a more comprehensive understanding of the topic under study, even if that specific content is not likely to be tested.

You will also find some review material from the AP Physics 1 course. This material is intended as a refresher, and is not a comprehensive review of pre-requisite content. It is anticipated that students using this book have a working understanding of AP Physics 1 material.

Table of Contents

Chapter 1: Introduction

"A theory that you can't explain to a bartender is probably no damn good."

— Ernest Rutherford

Objectives

Explore the scope of AP Physics 2 and review prerequisite skills for success.

1. Recognize the questions of physics.
2. List several disciplines within the study of physics.
3. Describe the key topics covered in AP Physics 2.
4. Define matter, mass, work and energy.
5. Explain how systems are comprised of constituent substructures.
6. Understand when a system may be treated as an object.
7. Recognize the intent and depth of this book as a companion resource to be used in conjunction with active learning practices such as hands-on exploration, discussion, debate, and deeper problem solving.

What is Physics?

Physics is many things to many different people. If you look up physics in the dictionary, you'll probably learn physics has to do with matter, energy, and their interactions. But what does that really mean? What is matter? What is energy? How do they interact? And most importantly, why do we care?

Physics, in many ways, is the answer to the favorite question of most 2-year-olds: "Why?" What comes after the why really doesn't matter. If it's a "why" question, chances are it's answered by physics: Why is the sky blue? Why does the wind blow? Why do we fall down? Why does my teacher smell funny? Why do airplanes fly? Why do the stars shine? Why do I have to eat my vegetables? The answer to all these questions, and many more, ultimately reside in the realm of physics.

Matter, Systems, and Mass

If physics is the study of matter, then we probably ought to define matter. **Matter**, in scientific terms, is anything that has mass and takes up space. In short, then, matter is anything you can touch – from objects smaller than electrons to stars hundreds of times larger than the sun. From this perspective, physics is the mother of all science. From astronomy to zoology, all other branches of science are subsets of physics, or specializations inside the larger discipline of physics.

In physics, you'll often times talk about **systems** as collections of smaller constituent substructures such as atoms and molecules. The properties and interactions of these substructures determine the properties of the system. In many cases, when the properties of the constituent substructures are not important in understanding the behavior of the system as a whole, the system is referred to as an **object**. For example, a baseball is comprised of many atoms and molecules which determine its properties. For the purpose of analyzing the path of a thrown ball in the air, however, the makeup of its constituent substructures is not important; therefore, we treat the ball as a single object.

So then, what is mass? **Mass** is, in simple terms, the amount of "stuff" an object is made up of. But of course, there's more to the story. Mass is split into two types: **inertial mass** and **gravitational mass**. Inertial mass is an object's resistance to being accelerated by a force. More massive objects accelerate less than smaller objects given an identical force. Gravitational mass, on the other hand, relates to the amount of gravitational force experienced by an object. Objects with larger gravitational mass experience a larger gravitational force.

Confusing? Don't worry! As it turns out, in all practicality, inertial mass and gravitational mass have always been equal for any object measured, even if it's not immediately obvious why this is the case (although with an advanced study of Einstein's Theory of General Relativity you can predict this outcome).

1.1 Q: On the surface of Earth, a spacecraft has a mass of 2.00×10^4 kg. What is the mass of the spacecraft at a distance of one Earth radius above Earth's surface?

(A) 5.00×10^3 kg

(B) 2.00×10^4 kg

(C) 4.90×10^4 kg

(D) 1.96×10^5 kg

1.1 A: (B) 2.00×10^4 kg. Mass is constant; therefore the spacecraft's mass at a distance of one Earth radius above Earth's surface is 2.00×10^4 kg.

Energy

If it's not matter, what's left? Why, energy, of course. As energy is such an everyday term that encompasses so many areas, an accurate definition can be quite elusive. Physics texts oftentimes define **energy** as the ability or capacity to do work. It's a nice, succinct definition, but leads to another question – what is work? **Work** can also be defined many ways, but a general definition starts with the process of moving an object. If you put these two definitions together, you can vaguely define energy as the ability or capacity to move an object.

Mass – Energy Equivalence

So far, the definition of physics boils down to the study of matter, energy, and their interactions. Around the turn of the 20th century, however, several physicists began proposing a strong relationship between matter and energy. Albert Einstein, in 1905, formalized this with his famous formula $E=mc^2$, which states that the mass of an object, a key characteristic of matter, is really a measure of its energy. This discovery has paved the way for tremendous innovation ranging from nuclear power plants to atomic weapons to particle colliders performing research on the origins of the universe. Ultimately, if traced back to its origin, the source of all energy on earth is the conversion of mass to energy!

Scope of AP Physics 2

Physics, in some sense, can therefore by defined as the study of just about everything. Try to think of something that isn't physics – go on, I dare you! Not so easy, is it? Even the more ambiguous topics can be categorized as physics. A Shakespearean sonnet? A sonnet is typically read from a manuscript (matter), sensed by the conversion of light (energy) alternately reflected and absorbed from a substrate, focused by a lens in the eye, and converted to chemical and electrical signals by photoreceptors on the retina. It is then transferred as electrical and chemical signals along the neural pathways to the brain. In short, just about everything is physics from a certain perspective.

As this book is focused on providing a resource to help with the College Board's AP Physics 2 curriculum, its scope is by necessity limited to topics covered by the AP Physics 2 curriculum. You'll begin with a study of fluids, including density, buoyancy, pressure, and the behaviors of moving fluids. Then you'll undertake a quick trip into the world of thermal physics, including temperature, heat, ideal gases, and basic thermodynamics.

Next, you'll revisit your AP-1 electricity studies with a review of AP-1 materials before going deeper into electric charges and forces, electric fields, electric potential difference, and basic parallel plate capacitors. You'll also have an opportunity to revisit circuits, dealing with more complex configurations of resistors as well as incorporating capacitors into your circuits. Your study of electricity will evolve as you explore magnetism and the relationship between electricity and magnetism. The next section of the book revisits basic wave characteristics and phenomena with a focus on electromagnetic waves, building a basic foundation in geometric and physical optics.

Finally, the AP-2 course will conclude with a number of topics under the heading of Modern Physics. You'll study wave-particle duality, models of the atom, radioactivity, mass-energy equivalence, and basic relativity.

The AP Physics 2 Exam itself is a three-hour exam designed to test your understanding of physics through a series of problems which require content knowledge, scientific inquiry skills, and the ability to reason logically and think critically. This is not an exam you can cram for, as it tests process skills such as your ability to apply physics concepts in unfamiliar contexts, explain physical relationships, design experiments, analyze data, and make generalizations from multiple areas of study. Instead, you should focus on building a strong understanding of the underlying principles emphasized in your course and in this book. To assist in this endeavor, the College Board has even published a list of seven "Big Ideas" that are prevalent throughout the study of physics, as well as associated "Enduring Understandings" that support these big ideas. These ideas and understandings will be emphasized throughout each chapter and highlighted in the chapter objectives.

7 Big Ideas in Physics

1. Objects and systems have properties such as mass and charge. Systems may have internal structure.
2. Fields existing in space can be used to explain interactions.
3. The interactions of an object with other objects can be described by forces.
4. Interactions between systems can result in changes in those systems.
5. Changes that occur as a result of interactions are constrained by conservation laws.
6. Waves can transfer energy and momentum from one location to another without the permanent transfer of mass and serve as a mathematical model for the description of other phenomena.
7. The mathematics of probability can be used to describe the behavior of complex systems and to interpret the behavior of quantum mechanical systems.

from the "AP Physics 1 & AP Physics 2 Curriculum Framework, 2014-2015"

In addition, the AP Physics course requires students to develop proficiency in seven "science practices" which are crucial to success. These practices, or skills, are best developed through an ongoing series of hands-on explorations and laboratory activities. As important as content knowledge is, physics is something you do, not just something you know. Therefore, the lab component is an important part of any AP Physics course.

7 Science Practices

1. Use representations and models to communicate and solve scientific problems.
2. Use mathematics appropriately.
3. Engage in scientific questioning to extend thinking or guide investigations.
4. Plan an experiment and collect data to answer a scientific question.
5. Analyze data and evaluate evidence.
6. Work with scientific explanations and theories.
7. Connect, relate, and apply knowledge across various concepts and models.

adapted from the "AP Physics 1 & AP Physics 2 Curriculum Framework, 2014-2015"

The test itself consists of two sections: a 90-minute multiple choice section and a 90-minute free response section. The multiple choice section consists of 50 to 55 questions with four answer choices per question. Unlike most multiple choice tests, however, certain questions may have multiple correct items that need to be chosen to receive full credit.

The free response section consists of five questions. Typically one question covers experimental design, one question covers quantitative and qualitative problem solving and reasoning, and three questions are of the short answer variety. It is expected that students are able to articulate their answers with a paragraph-length response. This paragraph should be organized, coherent, and make a logically reasoned argument with appropriate supporting data and evidence.

Note that many of this book's in-chapter questions are designed to assist you in solidifying the fundamental, underlying concepts required to succeed in the class. Many of these questions, therefore, are not at the level of complexity and integration across topics that you will see on the exam itself. They are designed to allow you to break the course into small, digestible chunks which you can then use to greater efficacy in labs, discussions, explorations, and deeper-understanding problems.

In preparing for the actual AP-2 exam itself, however, it is highly recommended that you practice a significant number of AP-2 style problems. You will find a large quantity of these types of problems (and answers) at the end of the book in the appendix. Use these problems to test your understanding and readiness for the exam itself. You can find additional AP-2 style problems from the APlusPhysics website as well as the College Board's AP Physics 2 website.

Chapter 2: Math Review

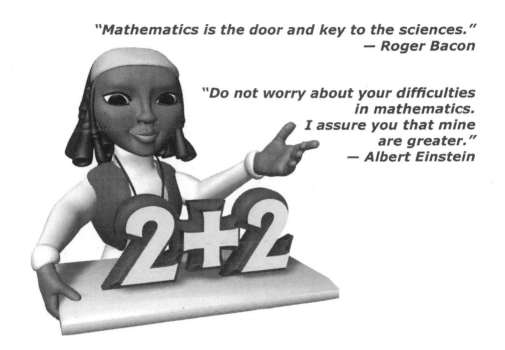

"Mathematics is the door and key to the sciences."
— *Roger Bacon*

*"Do not worry about your difficulties in mathematics.
I assure you that mine are greater."*
— *Albert Einstein*

Objectives

Review prerequisite math skills necessary for success.

1. Express answers correctly with respect to significant figures.
2. Use scientific notation to express physical values efficiently.
3. Convert and estimate SI units.
4. Differentiate between scalar and vector quantities.
5. Use scaled diagrams to represent and manipulate vectors.
6. Determine x- and y-components of two-dimensional vectors.
7. Determine the angle of a vector given its components.

Although physics and mathematics are not the same thing, they are in many ways closely related. Just like English is the language of this content, mathematics is the language of physics. A solid understanding of a few simple math concepts will allow you to communicate and describe the physical world both efficiently and accurately. Typically these concepts are developed in prior courses, though a refresher for some commonly used skills is provided in this chapter for your convenience.

Significant Figures

Significant Figures (or sig figs, for short) represent a manner of showing which digits in a number are known to some level of certainty. But how do you know which digits are significant? There are some rules to help with this. If you start with a number in scientific notation:

- All non-zero digits are significant.
- All digits between non-zero digits are significant.
- Zeroes to the left of significant digits are not significant.
- Zeroes to the right of significant digits are significant.

When you make a measurement in physics, you want to write what you measured using significant figures. To do this, write down as many digits as you are absolutely certain of, then take a shot at one more digit as accurately as you can. These are your significant figures.

2.1 Q: How many significant figures are in the value 43.74 km?

2.2 A: 4 (four non-zero digits)

2.2 Q: How many significant figures are in the value 4302.5 g?

2.2 A: 5 (All non-zero digits are significant and digits between non-zero digits are significant.)

2.3 Q: How many significant figures are in the value 0.0083s?

2.3 A: 2 (All non-zero digits are significant. Zeroes to the left of significant digits are not significant.)

2.4 Q: How many significant figures are in the value 1.200×10^3 kg?

2.4 A: 4 (Zeroes to the right of significant digits are significant.)

As the focus of this book is building a solid understanding of basic physics concepts and applications, significant figures will not be emphasized in routine problem solving, but realize that in certain environments they can be of the highest importance. For the purposes of the AP Physics 1 exam, typically 3-4 significant figures will be adequate.

Scientific Notation

Because measurements of the physical world vary so tremendously in size (imagine trying to describe the distance across the United States in units of hair thicknesses), scientists oftentimes use what is known as **scientific notation** to represent very large and very small numbers. These very large and very small numbers would become quite cumbersome to write out repeatedly. Imagine writing 4,000,000,000,000 over and over again. Your hand would get tired and your pen would rapidly run out of ink! Instead, it's much easier to write this number as 4×10^{12}. Or on the smaller scale, the thickness of the insulating layer (known as a gate dielectric) in the integrated circuits that power computers and other electronics can be less than 0.000000001 m. It's easy to lose track of how many zeros you have to deal with, so scientists instead would write this number as 1×10^{-9} m. See how much simpler life can be with scientific notation?

Scientific notation follows a few simple rules. Start by showing all the significant figures in the number you're describing, with the decimal point after the first significant digit. Then, show your number being multiplied by 10 to the appropriate power in order to give you the correct value.

It sounds more complicated than it is. Let's say, for instance, you want to show the number 300,000,000 in scientific notation (a very useful number in physics), and let's assume you know this value to three significant digits. You would start by writing the three significant digits, with the decimal point after the first digit, as "3.00". Now, you need to multiply this number by 10 to some power in order to get back to the original value. In this case, you multiply 3.00 by 10^8, for an answer of 3.00×10^8. Interestingly, the power you raise the 10 to is exactly equal to the number of digits you moved the decimal to the left as you converted from standard to scientific notation. Similarly, if you start in scientific notation, to convert to standard notation, all you have to do is remove the 10^8 power by moving the decimal point eight digits to the right. Presto, you're an expert in scientific notation!

But what do you do if the number is much smaller than one? Same basic idea. Let's assume you're dealing with the approximate radius of an electron, which is 0.00000000000000282 m. It's easy to see how unwieldy this could become. You can write this in scientific notation by writing out three significant digits, with the decimal point after the first digit, as "2.82." Again, you multiply this number by some power of 10 in order to get back to the original value. Because your value is less than 1, you need to use negative powers of 10. If you raise 10 to the power -15, specifically, you get a final value of 2.82×10^{-15} m. In essence, for every digit you moved the decimal place, you add another power of 10. And if you start with scientific notation, all you do is move the decimal place left one digit for every negative power of 10.

2.5 Q: Express the number 0.000470 in scientific notation.

2.5 A: 4.70×10^{-4}

2.6 Q: Express the number 2,870,000 in scientific notation.

2.6 A: 2.87×10^6

2.7 Q: Expand the number 9.56×10^{-3}.

2.7 A: 0.00956

2.8 Q: Expand the number 1.11×10^7.

2.8 A: 11,100,000

Metric System

Physics involves the study, prediction, and analysis of real-world phenomena. To communicate data accurately, you must set specific standards for basic measurements. The physics community has standardized what is known as the **Système International** (SI), which defines seven baseline measurements and their standard units, forming the foundation of what is called the metric system of measurement. The SI system is oftentimes referred to as the **mks system**, as the three most common measurement units are meters, kilograms, and seconds, which will be the focus for the majority of this course. The fourth SI base unit you'll use in this course, the ampere, will be introduced in the current electricity section.

The base unit of length in the metric system, the meter, is roughly equivalent to the English yard. For smaller measurements, the meter is divided up into 100 parts, known as centimeters, and each centimeter is made up of 10 millimeters. For larger measurements, the meter is grouped into larger units of 1000 meters, known as a kilometer. The length of a baseball bat is approximately one meter, the radius of a U.S. quarter is approximately a centimeter, and the diameter of the metal in a wire paperclip is roughly one millimeter.

The base unit of mass, the kilogram, is roughly equivalent to two U.S. pounds. A cube of water 10 cm x 10 cm x 10 cm has a mass of 1 kilogram. Kilograms can also be broken up into larger and smaller units, with commonly used measurements of grams (1/1000th of a kilogram) and milligrams (1/1000th of a gram). The mass of a textbook is approximately 2 to 3 kilograms, the mass of a baseball is approximately 145 grams, and the mass of a mosquito is 1 to 2 milligrams.

The base unit of time, the second, is likely already familiar. Time can also be broken up into smaller units such as milliseconds (10^{-3} seconds), microseconds (10^{-6} seconds), and nanoseconds (10^{-9} seconds), or grouped into larger units such as minutes (60 seconds), hours (60 minutes), days (24 hours), and years (365.25 days).

Chapter 2: Math Review

The metric system is based on powers of 10, allowing for easy conversion from one unit to another. A chart showing the meaning of commonly used metric prefixes and their notations can be extremely valuable in performing unit conversions.

Prefixes for Powers of 10		
Prefix	Symbol	Notation
tera	T	10^{12}
giga	G	10^9
mega	M	10^6
kilo	k	10^3
deci	d	10^{-1}
centi	c	10^{-2}
milli	m	10^{-3}
micro	μ	10^{-6}
nano	n	10^{-9}
pico	p	10^{-12}

Converting from one unit to another can be easily accomplished if you use the following procedure.

1. Write your initial measurement with units as a fraction over 1.
2. Multiply your initial fraction by a second fraction, with a numerator (top number) having the units you want to convert to, and the denominator (bottom number) having the units of your initial measurement.
3. For any units on the top right-hand side with a prefix, determine the value for that prefix. Write that prefix in the right-hand denominator. If there is no prefix, use 1.
4. For any units on the right-hand denominator with a prefix, write the value for that prefix in the right-hand numerator. If there is no prefix, use 1.
5. Multiply through the problem, taking care to accurately record units. You should be left with a final answer in the desired units.

Let's take a look at a sample unit conversion:

2.9 Q: Convert 23 millimeters (mm) to meters (m).

2.9 A: Step 1. $\dfrac{23mm}{1}$

Step 2. $\dfrac{23mm}{1} \times \dfrac{m}{mm}$

Step 3. $\dfrac{23mm}{1} \times \dfrac{m}{1mm}$

Step 4. $\dfrac{23mm}{1} \times \dfrac{10^{-3}m}{1mm}$

Step 5. $\dfrac{23mm}{1} \times \dfrac{10^{-3}m}{1mm} = 2.3 \times 10^{-2}m$

Now, try some on your own!

2.10 Q: Convert 2.67×10⁻⁴ m to mm.

2.10 A: $\dfrac{2.67 \times 10^{-4}m}{1} \times \dfrac{1mm}{10^{-3}m} = 0.267mm$

2.11 Q: Convert 14 kg to mg.

2.11 A: $\dfrac{14kg}{1} \times \dfrac{10^{3}mg}{10^{-3}kg} = 14 \times 10^{6}mg$

2.12 Q: Convert 3,470,000 μs to s.

2.12 A: $\dfrac{3,470,000\mu s}{1} \times \dfrac{10^{-6}s}{1\mu s} = 3.47s$

Accuracy and Precision

When making measurements of physical quantities, how close the measurement is to the actual value is known as the **accuracy** of the measurement. **Precision**, on the other hand, is the repeatability of a measurement. A common analogy involves an archer shooting arrows at the target. The bullseye of the target represents the actual value of the measurement.

| Low Accuracy | High Accuracy | Low Accuracy | High Accuracy |
| Low Precision | Low Precision | High Precision | High Precision |

Ideally, measurements in physics should be both accurate and precise.

Algebra and Trigonometry

Just as you find the English language a convenient tool for conveying your thoughts to others, you need a convenient language for conveying your understanding of the world around you in order to understand its behavior. The language most commonly (and conveniently) used to describe the natural world is mathematics. Therefore, to understand physics, you need to be fluent in the mathematics of the topics you'll study in this book - specifically basic algebra and trigonometry.

Now don't you fret or frown. You need only the most basic of algebra and trigonometry in order to successfully solve a wide range of physics problems.

A vast majority of problems requiring algebra can be solved using the same problem solving strategy. First, analyze the problem and write down what you know, what you need to find, and make a picture or diagram to better understand the problem if it makes sense. Then, start your solution by searching for a path that will lead you from your givens to your finds. Once you've determined an appropriate pathway (and there may be more than one), solve your problem algebraically for your answer. Finally, as your last steps, substitute in any values with units into your final equation, and solve for your answer, with units.

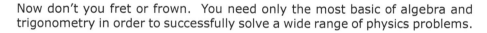

The use of trigonometry, the study of right triangles, can be distilled down to the definitions of the three basic trigonometric functions. When you know the length of two sides of a right triangle, or the length of one side and a non-right angle, you can solve for all the angles and sides of the triangle. If you can use the definitions of the sine, cosine, and tangent, you'll be fine in physics.

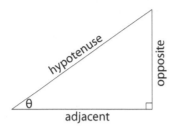

$$\sin \theta = \frac{opposite}{hypotenuse}$$

$$\cos \theta = \frac{adjacent}{hypotenuse}$$

$$\tan \theta = \frac{opposite}{adjacent}$$

Of course, if you need to solve for the angles themselves, you can use the inverse trigonometric functions (calculators should be in degree mode instead of radian mode).

$$\theta = \sin^{-1}\left(\frac{opposite}{hypotenuse}\right) = \cos^{-1}\left(\frac{adjacent}{hypotenuse}\right) = \tan^{-1}\left(\frac{opposite}{adjacent}\right)$$

2.13 Q: A car travels from the airport 14 miles east and 7 miles north to its destination. What direction should a helicopter fly from the airport to reach the same destination, traveling in a straight line?

2.13 A:

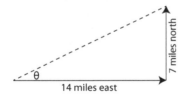

$$\theta = \tan^{-1}\left(\frac{opposite}{adjacent}\right)$$

$$\theta = \tan^{-1}\left(\frac{7 \text{ miles}}{14 \text{ miles}}\right) = 26.6°$$

2.14 Q: The sun creates a 10-meter-long shadow across the ground by striking a flagpole when the sun is 37 degrees above the horizon. How tall is the flagpole?

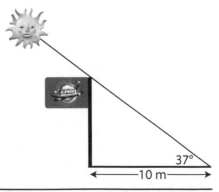

2.14 A: First, create a diagram of the situation, then recognize the shadow length is the adjacent side of the triangle, and the flagpole itself forms the opposite side. You can then use the tangent function to solve for the height of the flagpole.

$$\tan\theta = \frac{opp}{adj} \rightarrow opp = adj \times \tan\theta = (10m)\tan(37°) = 7.5m$$

Vectors and Scalars

Quantities in physics are used to represent real-world measurements, and therefore physicists use these quantities as tools to better understand the world. In examining these quantities, there are times when just a number, with a unit, can completely describe a situation. These numbers, which have just a **magnitude**, or size, are known as **scalars**. Examples of scalars include quantities such as temperature, mass, and time. At other times, a quantity is more descriptive if it also includes a direction. These quantities which have both a magnitude and direction are known as **vectors**. Vector quantities you may be familiar with include force, velocity, and acceleration.

Most students will be familiar with scalars, but to many, vectors may be a new and confusing concept. By learning just a few rules for dealing with vectors, you'll find that they can be a powerful tool for problem solving.

Vectors are often represented as arrows, with the length of the arrow indicating the magnitude of the quantity, and the direction of the arrow indicating the direction of the vector. Vectors out of the page are shown as dots, and vectors into the page are shown as X's. In the figure below, vector B has a magnitude greater than that of vector A even though vectors A and B point in the same direction. It's also important to know that vectors can be moved anywhere in space. The positions of A and B could be swapped, and the individual vectors would retain their values of magnitude and direction.

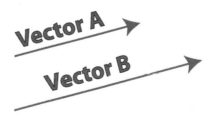

To add vectors A and B below, all you have to do is line them up so that the tip of the first vector touches the tail of the second vector. Then, to find the sum of the vectors, known as the **resultant**, draw a straight line from the start of the first vector to the end of the last vector. This method works with any number of vectors.

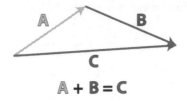

$$A + B = C$$

So how do you subtract two vectors? Try subtracting B from A. You can start by rewriting the expression A - B as A + -B. Now it becomes an addition problem. You just have to figure out how to express −B. This is easier than it sounds. To find the opposite of a vector, just point the vector in the opposite direction. Therefore, you can use what we already know about the addition of vectors to find the resultant of A-B.

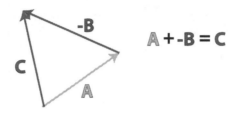

$$A + -B = C$$

Components of Vectors

You'll learn more about vectors as you go, but before moving on, there are a few basic skills to master. Vectors at angles can be challenging to deal with. By transforming a vector at an angle into two vectors, one parallel to the x-axis and one parallel to the y-axis, you can greatly simplify problem solving. To break a vector up into its components, you can use the basic trig functions.

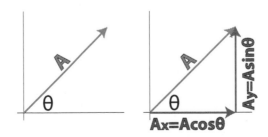

2.15 Q: The vector diagram below represents the horizontal component, F_H, and the vertical component, F_V, of a 24-newton force acting at 35° above the horizontal. What are the magnitudes of the horizontal and vertical components?

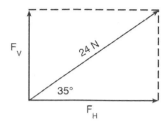

(A) F_H=3.5 N and F_V=4.9 N
(B) F_H=4.9 N and F_V=3.5 N
(C) F_H=14 N and F_V=20 N
(D) F_H=20 N and F_V=14 N

2.15 A: (D) $F_H = A_x = A\cos\theta = (24N)\cos 35° = 20N$

$F_V = A_Y = A\sin\theta = (24N)\sin 35° = 14N$

2.16 Q: An airplane flies with a velocity of 750 kilometers per hour, 30° south of east. What is the magnitude of the plane's eastward velocity?

(A) 866 km/h
(B) 650 km/h
(C) 433 km/h
(D) 375 km/h

2.16 A:

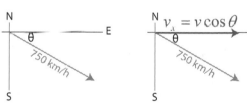

(B) $v_x = v\cos\theta = (750\,^{km}\!/_h)\cos(30°) = 650\,^{km}\!/_h$

2.17 Q: A soccer player kicks a ball with an initial velocity of 10 m/s at an angle of 30° above the horizontal. The magnitude of the horizontal component of the ball's velocity is

(A) 5.0 m/s
(B) 8.7 m/s
(C) 9.8 m/s
(D) 10 m/s

2.17 A:

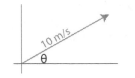

(B) $v_x = v\cos\theta = (10\,{}^m\!/_s)\cos(30°) = 8.7\,{}^m\!/_s$

2.18 Q: A child kicks a ball with an initial velocity of 8.5 meters per second at an angle of 35º with the horizontal, as shown. The ball has an initial vertical velocity of 4.9 meters per second. The horizontal component of the ball's initial velocity is approximately

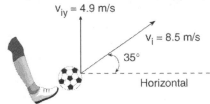

(A) 3.6 m/s
(B) 4.9 m/s
(C) 7.0 m/s
(D) 13 m/s

2.18 A: (C) $v_x = v\cos\theta = (8.5\,{}^m\!/_s)\cos(35°) = 6.96\,{}^m\!/_s$

In similar fashion, you can use the components of a vector in order to build the original vector. Graphically, if you line up the component vectors tip-to-tail, the original vector runs from the starting point of the first vector to the ending point of the last vector. To determine the magnitude of the resulting vector algebraically, just apply the Pythagorean Theorem.

2.19 Q: A motorboat, which has a speed of 5.0 meters per second in still water, is headed east as it crosses a river flowing south at 3.3 meters per second. What is the magnitude of the boat's resultant velocity with respect to the starting point?

(A) 3.3 m/s
(B) 5.0 m/s
(C) 6.0 m/s
(D) 8.3 m/s

2.19 A: (C) 6.0 m/s

The motorboat's resultant velocity is the vector sum of the motorboat's speed and the river's speed.

$$a^2 + b^2 = c^2$$
$$c = \sqrt{a^2 + b^2}$$
$$c = \sqrt{(5\tfrac{m}{s})^2 + (3.3\tfrac{m}{s})^2}$$
$$c = 6\tfrac{m}{s}$$

2.20 Q: A dog walks 8.0 meters due north and then 6.0 meters due east. Determine the magnitude of the dog's total displacement.

2.20 A: $a^2 + b^2 = c^2$
$$c = \sqrt{a^2 + b^2}$$
$$c = \sqrt{(6m)^2 + (8m)^2}$$
$$c = 10m$$

2.21 Q: A 5.0-newton force could have perpendicular components of

(A) 1.0 N and 4.0 N

(B) 2.0 N and 3.0 N

(C) 3.0 N and 4.0 N

(D) 5.0 N and 5.0 N

2.21 A: (C) The only answers that fit the Pythagorean Theorem are 3.0 N and 4.0 N ($3^2+4^2=5^2$)

2.22 Q: A vector makes an angle, θ, with the horizontal. The horizontal and vertical components of the vector will be equal in magnitude if angle θ is

(A) 30°

(B) 45°

(C) 60°

(D) 90°

2.22 A: (B) 45°. A_x=Acos(θ) will be equal to A_y=Asin(θ) when angle θ=45° since cos(45°)=sin(45°).

Vectors can also be added algebraically by adding their components to find the resultant vector.

2.23 Q: Kerbin the mouse travels 3 meters at an angle of 30 degrees north of east. He then travels 2 meters directly north. Finally, he travels 2 meters at an angle of 60 degrees south of west. What is the final position of Kerbin compared to his starting point?

2.23 A: You can first diagram this motion graphically, labeling the three different portions of the mouse's travel as vectors **A**, **B**, and **C**. By adding these three vectors to get resultant vector **R**, you determine Kerbin's final position. Start by lining up the three vectors tip-to-tail, drawing them to scale and using a protractor to make exact angles, then draw a vector from the starting point of the first vector to the ending point of the last vector to determine the resultant vector **R**.

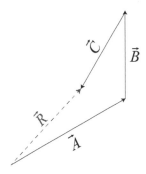

Alternately, you could break vectors **A**, **B**, and **C** into their components, then add up the individual x- and y-components to find the x- and y-components of resultant vector **R**.

$\vec{A} =< A\cos\theta, A\sin\theta >=< 3m\cos 30°, 3m\sin 30° >=< 2.60m, 1.5m >$

$\vec{B} =< B\cos\theta, B\sin\theta >=< 2m\cos 90°, 2m\sin 90° >=< 0m, 2m >$

$\vec{C} =< C\cos\theta, C\sin\theta >=< -2m\cos 60°, -2m\sin 60° >=< -1m, -1.73m >$

The resultant vector **R** is then found by adding up the x-components and y-components of the constituent vectors.

$R_X = A_X + B_X + C_X = 2.60m + 0m + -1m = 1.6m$

$R_Y = A_Y + B_Y + C_Y = 1.5m + 2m + -1.73m = 1.77m$

$\vec{R} =< R_X, R_Y >=< 1.6m, 1.77m >$

Therefore, the resultant position of Kerbin the mouse is 1.6m east and 1.77m north of his original starting position, which is 2.39 m from his starting position at an angle of 47.9 degrees north of east.

The Equilibrant Vector

The **equilibrant** of a force vector or set of force vectors is a single force vector which is exactly equal in magnitude and opposite in direction to the original vector or sum of vectors. The equilibrant, in effect, "cancels out" the original vector(s), or brings the set of vectors into equilibrium. To find an equilibrant, first find the resultant of the original vectors. The equilibrant is the opposite of the resultant you found!

2.24 Q: The diagram below represents two concurrent forces.

Which vector represents the force that will produce equilibrium with these two forces?

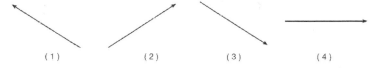

      (1)   (2)   (3)   (4)

2.24 A: (3) The resultant of the two vectors would point up and to the left; therefore the equilibrant must point in the opposite direction, down and to the right.

Other Pre-Requisite Skills

The AP Physics courses are advanced courses which, in reality, draw upon years and years of background skills such as effective writing, critical thinking, logical reasoning, graphing, and data analysis. These skills are typically developed by students across a multitide of prior courses, and should be reinforced regularly as part of lab activities, problem solving, discussion and debate.

Chapter 3: Fluids

"What is harder than rock,
or softer than water?
Yet soft water hollows out
hard rock.

Persevere."

— Ovid

Objectives

1. Calculate the density of an object.
2. Determine whether an object will float given its average density.
3. Calculate the forces on a submerged or partially submerged object using Archimedes' Principle.
4. Calculate pressure as the force a system exerts over an area.
5. Explain the operation of a hydraulic system as a function of equal pressure throughout a fluid.
6. Apply the continuity equation to fluids in motion.
7. Apply Bernoulli's Principle to fluids in motion.

If you're going to take a whole chapter to study fluids, it would make sense to start with what fluids are. A **fluid** is matter that flows under pressure, which includes liquids, gases, and even plasmas. Water is a fluid, air is a fluid, the sun is a fluid, even molasses is a fluid. **Fluid Mechanics** is the study of fluids, ranging from fluids at rest, to fluids in motion, to forces applied to and exerted by fluids. You could start your study of fluids in a variety of places, but one of the simplest examples of fluid behavior comes from investigating objects that float and objects that sink. To understand this behavior, why not begin with density?

Density

Density is defined as the ratio of an object's mass to the volume it occupies, and is frequently given the symbol rho (ρ) in physics.

$$\rho = \frac{m}{V}$$

Less dense fluids will float on top of more dense fluids, and less dense solids will float on top of more dense fluids (keeping in mind you must look at the average density of the entire solid object).

3.01 Q: A single kilogram of water fills a cube of length 0.1m. What is the density of water?

3.01 A: $\rho = \frac{m}{V}$

$$\rho = \frac{1kg}{(0.1m)(0.1m)(0.1m)} = 1000 \, ^{kg}\!/_{m^3}$$

3.02 Q: Gold has a density of 19,320 kg/m³. How much volume does a single kilogram of gold occupy?

3.02 A: $\rho = \frac{m}{V}$

$$V = \frac{m}{\rho} = \frac{1kg}{19320 \, ^{kg}\!/_{m^3}} = 5.18 \times 10^{-5} m^3$$

3.03 Q: Fresh water has a density of 1000 kg/m³. Which of the following materials will float on water?

(A) Ice (ρ=917 kg/m³)

(B) Magnesium (ρ=1740 kg/m³)

(C) Cork (ρ=250 kg/m³)

(D) Glycerol (ρ=1261 kg/m³)

3.03 A: (A) and (C). Both ice and cork will float on water because they have an average density less than that of water.

3.04 Q: Based on the image below, what can you say about the average density of the man and inner tube compared to the density of the water?

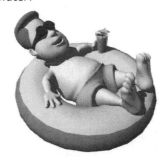

(A) The average density of the man and inner tube is greater than that of the water.

(B) The average density of the man and inner tube is less than that of the water.

(C) The average density of the man and inner tube is equal to that of the water.

3.04 A: (B) The average density of a solid must be less than that of any fluid it is floating in.

3.05 Q: Four spheres of uniform density are designed such that spheres A and B are made of the same material, and spheres C and D are made of the same material. The mass of sphere A is equal to the mass of sphere D. Given the diagram below, is the mass of sphere B greater than, less than, or equal to the mass of sphere C?

3.05 A: The mass of sphere B is greater than the mass of sphere C. Because the mass of sphere A is equal to the mass of sphere D, you can state that the density of spheres A and B is greater than the density of spheres C and D. Since sphere B and sphere C are the same size, and sphere B has a higher density than sphere C, sphere B must have a greater mass than sphere C.

Buoyancy

As you can imagine, there is definitely more to whether an object floats or not than just average density. For example, why do some objects float higher in the water than others? And why is it easier to lift objects underwater than in the air? To answer these questions, you'll need to understand the concept of **buoyancy**, a force which is exerted by a fluid on an object, opposing the object's weight.

It is rumored that the Greek philosopher and scientist Archimedes, around 250 B.C., was asked by King Hiero II to help with a problem. King Hiero II had ordered a fancy golden crown from a goldsmith. However, the king was concerned that the goldsmith may have taken his money and mixed some silver in with the crown instead of crafting the crown out of pure gold. He asked Archimedes if there was a way to determine if the crown was pure gold.

Archimedes puzzled over the problem for some time, coming up with the solution while he was in the bathtub one evening. When Archimedes submerged himself in the tub, he noticed that the amount of water that spilled over the rim of the tub was equal to the volume of water he displaced.

Using this method, he could place the crown in a bowl full of water. The amount of water that spilled over could be measured and used to tell the volume of the crown. By then dividing the mass of the crown by the volume, he could obtain the density of the crown, and compare the density to that of gold, determining if the crown was pure gold. According to legend, he was so excited he popped out of the tub and ran through the streets naked yelling "Eureka! Eureka!" (Greek for "I found it! I found it!").

True story or not, this amusing tale illustrates Archimedes' development of a key principle of buoyancy: the buoyant force (F_B) on an object is equal to the density of the fluid, multiplied by the volume of the fluid displaced (which is also equal to the volume of the submerged portion of the object), multiplied by the gravitational field strength. This is known as **Archimedes' Principle**.

                                                       Chapter 3: Fluids

$$F_B = \rho_{fluid} V g$$

Archimedes' Principle explains why boats made of steel can float. Although the steel of the boat itself is more dense than water, the average density of the entire boat (including the air in the interior of the boat) is less than that of water. Put another way, the boat floats because the weight of the volume of water displaced by the boat is greater than the weight of the boat itself.

This principle also accounts for the ability of submarines to control their depth. Submarines use pumps to move water into and out of chambers in their interior, effectively controlling the average density of the submarine. If the submarine wants to rise, it pumps water out, reducing its average density. If it wants to submerge, it pumps water in, increasing its average density.

3.06 Q: What is the buoyant force on a 0.3 m³ box which is fully submerged in freshwater (density=1000 kg/m³)?

3.06 A: $F_B = \rho_{fluid} V g = (1000\,{}^{kg}\!/_{m^3})(0.3m^3)(9.8\,{}^{m}\!/_{s}) = 2940N$

3.07 Q: A steel cable holds a 120-kg shark tank 3 meters below the surface of saltwater. If the volume of water displaced by the shark tank is 0.1 m³, what is the tension in the cable? Assume the density of saltwater is 1025 kg/m³.

3.07 A: First, draw a free body diagram (FBD) of the situation, realizing that you have the force of gravity (mg) pulling down, the buoyant force upward, and the force of tension in the cable upward.

Because the shark tank is at equilibrium under the water, the net force on it must be zero, therefore the upward forces must balance the downward forces. You can write this using Newton's 2nd Law in the y-direction as:

$$F_{NET_y} = F_T + F_B - mg = 0$$

Finally, you can use this equation to solve for the force of tension in the cable.

$$F_T = mg - F_B$$
$$F_T = mg - \rho_{fluid}Vg$$
$$F_T = (120kg)(9.8\,m/_{s^2}) - (1025\,kg/_{m^3})(0.1m^3)(9.8\,m/_{s^2})$$
$$F_T = 172N$$

3.08 Q: A rectangular boat made out of concrete with a mass of 3000 kg floats on a freshwater lake (ρ=1000 kg/m³). If the bottom area of the boat is 6 m², how much of the boat is submerged?

3.08 A: Because the boat is floating on the lake, the magnitude of the buoyant force must be equal to the magnitude of the weight of the boat (F_B=mg).

Since the boat is rectangular, you can write its volume (V) as its bottom area (A=6 m²) multiplied by the depth submerged (d).

$$F_B = mg$$
$$\rho_{fluid}Vg = mg$$
$$\rho_{fluid}(Ad)g = mg$$
$$d = \frac{m}{\rho_{fluid}A} = \frac{3000kg}{(1000\,kg/_{m^3})(6m^2)} = 0.5m$$

3.09 Q: Four spheres of identical size are suspended in identical cups of water by a string as shown in the diagram below. Each sphere has a different mass and is completely submerged to the same depth. Rank the buoyant force on the blocks from largest to smallest

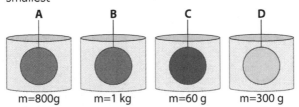

A   B   C   D

m=800g m=1 kg m=60 g m=300 g

3.09 A: A=B=C=D. The buoyant force on the spheres is a function of their volume and the density of the fluid displaced. All spheres have the same volume, displace the same amount of fluid, and displace fluid of the same density (water); therefore they are all subjected to the same buoyant force.

Pressure

Everyone's been under pressure at one time or another, or in certain circumstances really "felt the pressure." From a scientific perspective, however, pressure has a very specific definition, and its exploration leads to some very important applications.

In physics, **pressure** is the effect of a force acting upon a surface. Mathematically, it is a scalar quantity calculated as the force applied per unit area, where the force applied is always perpendicular to the surface. The SI unit of pressure, a Pascal (Pa), is equivalent to a N/m².

$$P = \frac{F}{A}$$

All states of matter can exert pressure. When you walk across an ice-covered lake, you are applying a pressure to the ice equal to the force of gravity on your body (your weight) divided by the area over which you're contacting the ice. This is why it is important to spread your weight out when traversing fragile surfaces. Your odds of breaking through the ice go up tremendously if you walk across the ice in high heels, as the small area contacting the ice leads to a high pressure. This is also the reason snow shoes have such a large area. They are designed to reduce the pressure applied to the top crust of snow so that you can walk more easily without sinking into snow drifts.

Fluids can also exert pressure. All fluids exert outward pressure in all directions on the sides of any container holding the fluid. Even the Earth's atmosphere exerts pressure, which you are experiencing right now. The pressures inside and outside your body are so well balanced, however, that you rarely notice the 101,325 Pascals due to the atmosphere (approximately 10N/cm²). If you ride up or down a steep hill in a car and change altitude (and therefore pressure) quickly, you may have experienced a "popping" sensation in your ears — this is due to the pressure inside your ear balancing the pressure outside your ear in a transfer of air through small tubes that connect your inner ear to your throat.

3.10 Q: Air pressure is approximately 100,000 Pascals. What force is exerted on the top surface of this book when it is sitting flat on a desk? The area of the book's cover is 0.035 m².

3.10 A: $P = \frac{F}{A}$

$F = PA = (100,000\,Pa)(0.035m^2) = 3500N$

3.11 Q: A fisherman with a mass of 75-kg falls asleep on his four-legged chair of mass 5 kg. If each leg of the chair has a surface area of 2.5×10^{-4} m² in contact with the ground, what is the average pressure exerted by the fisherman and chair on the ground?

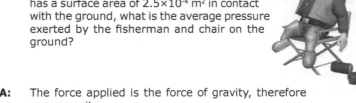

3.11 A: The force applied is the force of gravity, therefore we can write:

$$P = \frac{F}{A} = \frac{mg}{A} = \frac{(75kg + 5kg)(9.8\,^{m}/_{s})}{4(2.5 \times 10^{-4}\,m^{2})} = 784,000\,Pa$$

3.12 Q: A device which measures pressure on a surface reads 0 in the vacuum of space. It is then placed on the surface of planet Physica. On the planet's surface, the device indicates a force of 10,000 Newtons on its surface. Calculate the surface area of the device, given the pressure on the surface is 80,000 Pascals.

3.12 A: $$P = \frac{F}{A}$$

$$A = \frac{F}{P} = \frac{10,000N}{80,000Pa} = 0.125m^{2}$$

3.13 Q: Rank the following from highest pressure to lowest pressure upon the ground:

(A) The atmosphere at sea level

(B) A 7000-kg elephant with total area 0.5 m² in contact with the ground

(C) A 65-kg lady in high heels with total area 0.005 m² in contact with the ground

(D) A 1600-kg car with a total tire contact area of 0.2 m²

3.13 A: (B) The elephant (137,000 Pa)

(C) The lady in high heels (127,000 Pa)

(A) The atmosphere (100,000 Pa)

(D) The car (78,400 Pa)

The pressure that a fluid exerts on an object submerged in that fluid can be calculated almost as simply. If the object is submersed to a depth (h), the pressure is found by multiplying the density of the fluid by the depth submerged, all multiplied by the acceleration due to gravity.

$$P_{gauge} = \rho g h$$

This is known as the gauge pressure, because this is the reading you would observe on a pressure gauge. If there is also atmosphere above the fluid, such as the situation here on earth, you can determine the absolute pressure, or total pressure, by adding in the atmospheric pressure (P_0), which is equal to approximately 100,000 Pascals.

$$P_{absolute} = P_0 + P_{gauge} = P_0 + \rho g h$$

3.14 Q: Samantha spots buried treasure while scuba diving on her Caribbean vacation. If she must descend to a depth of 40 meters to examine the treasure, what gauge pressure will she read on her scuba equipment? The density of sea water is 1025 kg/m³.

3.14 A: $P_{gauge} = \rho g h$

$P_{gauge} = (1025\,{}^{kg}\!/_{m^3})(9.8\,{}^{m}\!/_{s^2})(40m) = 402,000\,Pa$

3.15 Q: What is the absolute pressure exerted on the diver in the previous problem by the water and atmosphere?

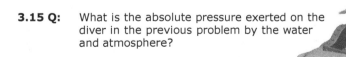

3.15 A: $P_{absolute} = P_0 + P_{gauge}$

$P_{absolute} = 100,000\,Pa + 400,000\,Pa = 500,000\,Pa$

3.16 Q: A diver's pressure gauge reads 250,000 Pascals in fresh water (ρ=1000 kg/m³). How deep is the diver?

3.16 A: $P = \rho g h$

$$h = \frac{P}{\rho g} = \frac{250,000\,Pa}{(1000\,^{kg}/_{m^3})(9.8\,^{m}/_{s^2})} = 25.5m$$

3.17 Q: Identical pressure gauges are placed in four cylinders of various size and shape. Each cylinder is filled with water, and each pressure gauge is the same distance from the bottom of the cylinder. Rank the pressure on each pressure gauge from greatest to least.

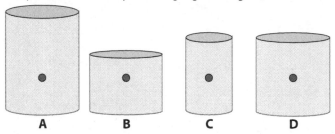

A   **B**   **C**   **D**

3.17 A: A, C=D, B. The pressure on each gauge is a function of the depth to which it is submerged.

Pascal's Principle

When a force is applied to a contained, incompressible fluid, the pressure increases equally in all directions throughout the fluid. This fundamental characteristic of fluids provides the foundation for hydraulic systems found in barbershop chairs, construction equipment, and the brakes in your car.

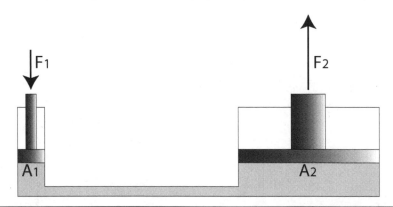

Because the force applied to the contained fluid is distributed throughout the system, you can multiply the applied force through this application of Pascal's Principle in the following manner. Assume you have a closed container filled with an incompressible fluid with two pistons of differing areas, A_1 and A_2. If you apply a force, F_1, to the piston of area A_1, you create a pressure in the fluid which you can call P_1.

$$P_1 = \frac{F_1}{A_1}$$

Similarly, the pressure at the second piston, P_2, must be equal to F_2 divided by the area of the second piston, A_2.

$$P_2 = \frac{F_2}{A_2}$$

Since the pressure is transmitted equally throughout the fluid In all directions according to Pascal's Principle, P_1 must equal P_2.

$$P_1 = P_2 \rightarrow \frac{F_1}{A_1} = \frac{F_2}{A_2}$$

Rearranging to solve for F_2, you find that F_2 is increased by the ratio of the areas A_2 over A_1.

$$F_2 = \frac{A_2}{A_1} F_1$$

Therefore, you have effectively increased the applied force F_1. Of course, the law of conservation of energy cannot be violated, so the work done on the system must balance the work done by the system. In the hydraulic lift diagram shown on the previous page, the distance over which F_1 is applied will be greater than the distance over which F_2 is applied, by the exact same ratio as the force multiplier!

3.18 Q: A barber raises his customer's chair by applying a force of 150N to a hydraulic piston of area 0.01 m². If the chair is attached to a piston of area 0.1 m², how massive a customer can the chair raise? Assume the chair itself has a mass of 5 kg.

3.18 A: To solve this problem, first determine the force applied to the larger piston.

$$F_2 = \frac{A_2}{A_1} F_1$$

$$F_2 = \frac{0.10m^2}{0.01m^2} (150N) = 1500N$$

If the maximum force on the chair is 1500N, you can now determine the maximum mass which can be lifted by recognizing that the force that must be overcome to lift the customer is the force of gravity; therefore the applied force on the customer must equal the force of gravity on the customer.

$$F = mg$$

$$m = \frac{F}{g} = \frac{1500N}{9.8\,^m/_{s^2}} = 153kg$$

If the chair has a mass of 5 kilograms, the maximum mass of a customer in the chair must be 148 kg.

3.19 Q: A hydraulic system is used to lift a 2000-kg vehicle in an auto garage. If the vehicle sits on a piston of area 0.5 square meters, and a force is applied to a piston of area 0.03 square meters, what is the minimum force that must be applied to lift the vehicle?

(A) 11,600 N

(B) 3330 N

(C) 1180 N

(D) 120 N

3.19 A: (C) 1180 N

$$P_1 = P_2 \rightarrow \frac{F_1}{A_1} = \frac{F_2}{A_2}$$

$$F_1 = \frac{A_1}{A_2} F_2 = \frac{0.03m^2}{0.5m^2} \left(2000kg \times 9.8\,^m/_{s^2} \right)$$

$$F_1 = 1180N$$

Continuity Equation for Fluids

When fluids move through a full pipe, the volume of fluid that enters the pipe must equal the volume of fluid that leaves the pipe, even if the diameter of the pipe changes. This is a restatement of the law of conservation of mass for fluids.

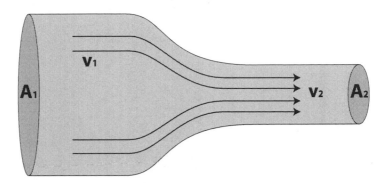

The volume of fluid moving through the pipe at any point can be quantified in terms of the volume flow rate, which is equal to the area of the pipe at that point multiplied by the velocity of the fluid. This volume flow rate must be constant throughout the pipe; therefore you can write the equation of continuity for fluids (also known as the fluid continuity equation) as:

$$A_1 v_1 = A_2 v_2$$

This equation says that as the cross-section of the pipe gets smaller, the velocity of the fluid increases, and as the cross-section gets larger, the fluid velocity decreases. You may have applied this yourself in watering the flowers with a garden hose. If you want increase the velocity of the water coming from the end of the hose, you place your thumb over part of the opening of the hose, effectively decreasing the cross-sectional area of the hose's end and increasing the velocity of the exiting water!

3.20 Q: Water runs through a water main of cross-sectional area 0.4 m² with a velocity of 6 m/s. Calculate the velocity of the water in the pipe when the pipe tapers down to a cross-sectional area of 0.3 m².

(A) 4.5 m/s

(B) 6 m/s

(C) 8 m/s

(D) 10.7 m/s

3.20 A: (C) $A_1 v_1 = A_2 v_2$

$$v_2 = \frac{A_1}{A_2} v_1 = \frac{0.4 m^2}{0.3 m^2} (6 \, m/s) = 8 \, m/s$$

3.21 Q: Water enters a typical garden hose of diameter 1.6 cm with a velocity of 3 m/s. Calculate the exit velocity of water from the garden hose when a nozzle of diameter 0.5 cm is attached to the end of the hose.

3.21 A: First, find the cross-sectional areas of the entry (A_1) and exit (A_2) sides of the hose.

$$A_1 = \pi r^2 = \pi(0.008m)^2 = 2 \times 10^{-4}\, m^2$$

$$A_2 = \pi r^2 = \pi(0.0025m)^2 = 1.96 \times 10^{-5}\, m^2$$

Next, apply the continuity equation for fluids to solve for the water velocity as it exits the hose (v_2).

$$A_1 v_1 = A_2 v_2$$

$$v_2 = \frac{A_1}{A_2}\, v_1 = \frac{2 \times 10^{-4}\, m^2}{1.96 \times 10^{-5}\, m^2}\,(3\,^m/_s) = 30.6\,^m/_s$$

Bernoulli's Principle

Conservation of energy, when applied to fluids in motion, leads to Bernoulli's Principle. **Bernoulli's Principle** states that fluids moving at higher velocities lead to lower pressures, and fluids moving at lower velocities result in higher pressures.

Airplane wings have a larger top surface than a bottom surface to take advantage of this fact. As the air moves across the larger top surface, it must move faster than the air traveling a shorter distance under the bottom surface. This leads to a lower pressure on top of the wing, and a higher pressure underneath the wing, providing some of the lift for the aircraft (note that this isn't the only cause of lift, as Newton's 3rd Law also plays a critical role in understanding the dynamics of flight).

This principle is also used in sailboats, carburetors, gas delivery systems, and even water-powered sump pumps!

Chapter 3: Fluids

Expressing Bernoulli's Principle quantitatively, you can relate the pressure, velocity, and height of a liquid in a tube at various points.

$$P_1 + \tfrac{1}{2}\rho v_1^2 + \rho g y_1 = P_2 + \tfrac{1}{2}\rho v_2^2 + \rho g y_2$$

The pressure at a point in the tube plus half the density of the fluid multiplied by the square of its velocity at that point, added to ρgy, must be equal at any point in the tube.

3.22 Q: Water sits in a large open jug at a height of 0.2m above the spigot. With what velocity will the water leave the spigot when the spigot is opened?

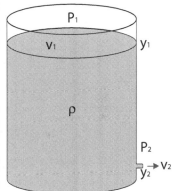

3.22 A: Since the top of the jug and the spigot are both open to atmosphere, the pressures P_1 and P_2 must be equal. Since the jug is much larger than the spigot, you can assume the velocity of the water at the top of the jug is nearly zero. This allows you to simplify Bernoulli's Equation considerably.

$$P_1 + \tfrac{1}{2}\rho v_1^2 + \rho g y_1 = P_2 + \tfrac{1}{2}\rho v_2^2 + \rho g y_2$$

$$\rho g y_1 = \tfrac{1}{2}\rho v_2^2 + \rho g y_2$$

Since the density of the fluid is the same throughout, you can do some algebraic simplification to solve for v_2.

$$g(y_1 - y_2) = \tfrac{1}{2}v_2^2$$

$$v_2 = \sqrt{2g(y_1 - y_2)}$$

This is known as Torricelli's Theorem. Since the difference in height is 0.2m, you can now easily solve for the velocity of the water at the spigot.

$$v_2 = \sqrt{2g(y_1 - y_2)}$$

$$v_2 = \sqrt{2(9.8\,\tfrac{m}{s^2})(0.2m)} = 1.98\,\tfrac{m}{s}$$

Notice that this is the same result you would obtain if you had solved for the velocity of an object dropped from a height of 0.2 meters using the kinematic equations... this should make sense, as Bernoulli's Equation is really just a restatement of conservation of energy, applied to fluids!

3.23 Q: Water leaves a hot water heater in a basement through a pipe of radius 2.5 cm at a speed of 1 m/s under a pressure of 200 kPa. What is the speed and pressure of the water in the bedroom 6 meters above, where the pipe has narrowed to a 2-cm diameter?

3.23 A: Utilize the continuity equation to find the speed of the water in the bedroom.

$$A_1 v_1 = A_2 v_2 \rightarrow v_2 = \frac{A_1 v_1}{A_2} = \frac{\pi(0.025m)^2(1\,m/s)}{\pi(0.01m)^2} = 6.25\,m/s$$

Then, Bernoulli's Equation can be applied to determine the pressure in the bedroom.

$$P_1 + \tfrac{1}{2}\rho v_1^2 + \rho g y_1 = P_2 + \tfrac{1}{2}\rho v_2^2 + \rho g y_2 \xrightarrow{y_1=0} P_2 = P_1 + \tfrac{1}{2}\rho(v_1^2 - v_2^2) - \rho g y_2 \rightarrow$$
$$P_2 = 200000 + \tfrac{1}{2}(1000)(1^2 - 6.25^2) - (1000)(9.8)(6) \approx 122000\,Pa = 122kPa$$

Test Your Understanding

1. The phrase "tip of the iceberg" is often used to describe a phenomena in which just a small portion of the greater whole is visible or obvious. Explain why only a small portion of an iceberg floats above the surface, while a majority of the iceberg is submerged.

2. Determine the density of helium by measuring the forces on a floating helium balloon held in place by a string.

3. A jar of spaghetti sauce is sealed under partial vacuum with a metal lid. In the center of the metal lid is a small button and the words "Safety button pops up when original seal is broken." Explain why the button remains in while the jar is sealed.

4. While drinking a soda from a straw, you dip the straw into the liquid, and place your finger on top of the straw. Explain why the liquid remains in the straw when you remove the straw from the liquid.

5. Examine Bernoulli's Equation in detail, recognizing it as a restatement of conservation of energy in a moving fluid. Multiply both sides of the equation by volume (V). How do the various expressions in the equation relate to expressions for energy you've studied previously? Explain.

6. Physicists have long debated what the driving factor is in an airplane wing creating lift -- the Bernoulli Effect due to the air's longer path above the wing, or Newton's 3rd Law with the wing deflecting air partially downward, therefore the air pushes upward on the wing. Stage a debate in your class, justifying your claims using fundamental physics principles.

Chapter 4: Thermal Physics

"Thermodynamics is a funny subject.
The first time you go through it,
you don't understand it at all.

The second time you go through it,
you think you understand it,
except for one or two small points.

The third time you go through it,
you know you don't understand it,
but by that time you are so used to it,
it doesn't bother you anymore."

— Arnold Sommerfeld

Objectives

1. Calculate the temperature of an object given its average kinetic energy.
2. Calculate the linear and volumetric expansion of a solid as a function of temperature.
3. Explain heat as the process of transferring energy between systems at different temperatures.
4. Calculate an object's temperature change using its specific heat.
5. Determine the energy required for a material to undergo a phase change.
6. Utilize the ideal gas law to solve for pressure, volume, temperature, and quantity of an ideal gas.
7. Describe the zeroth, first, second, and third laws of thermodynamics.
8. Utilize PV diagrams to describe changes in ideal gas conditions.
9. Analyze adiabatic, isobaric, isochoric, and isothermal processes using both algebraic and graphical methods.

Thermal physics deals with the internal energy of objects due to the motion of the atoms and molecules comprising the objects, as well as the transfer of this energy from object to object, known as heat.

Temperature

The internal energy of an object, known as its thermal energy, is related to the kinetic energy of all the particles comprising the object. The more kinetic energy the constituent particles have, the greater the object's thermal energy.

In solids, the particles comprising the solid are held together tightly; therefore their motion is limited to vibrating back and forth in their given positions. In liquids, the particles can move back and forth across each other, but the object itself has no defined shape. In gases, the particles move throughout the volume available, interacting with each other and the walls of any container holding them. In all cases, the total thermal energy of the object is the sum total of the kinetic energies of its constituent particles.

Instead of just looking at the sum of all the individual particles' kinetic energies, you could examine the average kinetic energy of the particles comprising the object, realizing that the actual kinetic energies of individual particles may vary significantly and can be modeled as a statistical distribution. The average kinetic energy of the particles is directly related to the temperature of the object by the following equation:

$$K_{avg} = \frac{3}{2} k_B T$$

Examining this equation, the average kinetic energy is given in Joules, k_B is Boltzmann's Constant (1.38×10^{-23} J/K), and the temperature is given in Kelvins, the SI unit of temperature. Note that even though two objects can have the same temperature (and therefore the same average kinetic energy), they may have different internal energies.

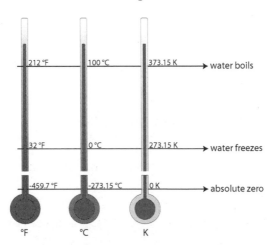

The Kelvin scale is closely related to the Celsius temperature scale, but where the Celsius scale targets the freezing point of water as 0° C, the Kelvin scale utilizes its zero at what is known as absolute zero (the point on a Volume vs. Temperature graph for a gas where the extended curve would hypothetically reach zero volume), considered a theoretical minimum temperature. Therefore, absolute zero is 0 Kelvins, which is equivalent to -273.15° Celsius. To convert from Kelvins to degrees Celsius, just add 273.15 to your temperature reading in degrees Celsius. The freezing point of water, then, is 0°C or 273.15 K, and the boiling point of water is 100°C or 373.15 K. Compare this to the Fahrenheit scale, where water freezes at 32°F, and boils at 212°F!

$$T_K = T_{°C} + 273.15$$
$$T_{°F} = \tfrac{9}{5}T_{°C} + 32$$
$$T_{°C} = \tfrac{5}{9}(T_{°F} - 32)$$

4.01 Q: What is the average kinetic energy of the molecules in a steak at a temperature of 345 Kelvins?

(A) 223 J

(B) 4.76×10⁻²¹ J

(C) 7.14×10⁻²¹ J

(D) 517 J

4.01 A: (C) $K_{avg} = \dfrac{3}{2}k_B T = \dfrac{3}{2}(1.38 \times 10^{-23}\ \text{J/}_K)(345K) = 7.14 \times 10^{-21}\ J$

4.02 Q: Normal canine body temperature is 101.5°F. What is normal canine body temperature in degrees Celsius? In Kelvins?

4.02 A: $T_{°C} = \tfrac{5}{9}(T_{°F} - 32) = \tfrac{5}{9}(101.5 - 32) = 38.6°C$
$T_K = T_{°C} + 273.15 = 38.6 + 273.15 = 311.75K$

4.03 Q: The average temperature of space is estimated as roughly -270.4°C. What is the average kinetic energy of the particles in space?

4.03 A: First, convert the temperature into Kelvins.

$$T_K = T_{°C} + 273.15 = -270.4 + 273.15 = 2.75K$$

Next, solve for the average kinetic energy of the particles.

$$K_{avg} = \frac{3}{2}k_B T = \frac{3}{2}(1.38 \times 10^{-23} \, J/_K)(2.75K) = 5.69 \times 10^{-23} J$$

4.04 Q: Given that the average kinetic energy of the particles comprising our sun is 1.2×10⁻¹⁹ J, find the temperature of the sun.

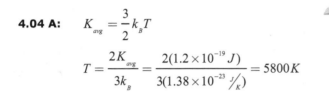

4.04 A: $$K_{avg} = \frac{3}{2}k_B T$$

$$T = \frac{2K_{avg}}{3k_B} = \frac{2(1.2 \times 10^{-19} J)}{3(1.38 \times 10^{-23} \, J/_K)} = 5800K$$

4.05 Q: Which graph best represents the relationship between the average kinetic energy (K_{avg}) of the random motion of the molecules of an ideal gas and its absolute temperature (T)?

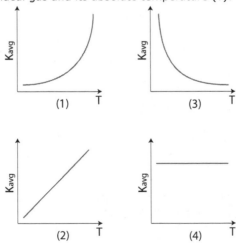

4.05 A: (2) since K_{avg} is a linear function of absolute temperature T.

4.06 Q: While orbiting Earth, the space shuttle has recorded temperatures ranging from 398K to 118K. These temperatures correspond to Celsius temperatures ranging from

(A) 125°C to -391°C

(B) 125°C to -155°C

(C) 671°C to 391°C

(D) 671°C to 155°C

4.06 A: (B) 125°C to -155°C

4.07 Q: The temperature of an object which cannot transfer thermal energy to another object is

(A) -273 K

(B) 0 K

(C) 0°C

(D) 273°C

4.07 A: (B) 0 K is absolute zero.

4.08 Q: Rank from lowest to highest the temperatures T=100°F, T=100°C, and T=100 K.

4.08 A: 100 K, 100°F, 100°C

Thermal Expansion

When objects are heated, they tend to expand, and when they are cooled, they tend to contract. You can use this to open glass jars with tight metal lids by running the lids under hot water. The temperature increase in the lid expands the metal lid and the glass jar, but because most metals expand more quickly than glass, the lid becomes looser, making it easier to open the jar.

This occurs because at higher temperatures, objects have higher kinetic energies, so their particles vibrate more. At these higher levels of vibration, the particles aren't bound to each other as tightly, so the object expands.

The amount an object expands can be calculated for both one-dimensional (linear) and three-dimensional (volumetric) expansion. The amount a material expands is characterized by the material's coefficient of expansion. In calculating a material's one-dimensional expansion, you can use the linear expansion formula, and the material's linear coefficient of expansion (α).

$$\Delta l = \alpha l_0 \Delta T$$

When calculating a material's three-dimensional expansion, you'll use the volumetric expansion formula, and the material's volumetric coefficient of expansion (β). Note that in most cases, the volumetric coefficient of expansion is roughly three times the linear coefficient of expansion.

$$\Delta V = \beta V_0 \Delta T$$

The change in temperature in the expansion equations can be given in either degrees Celsius or Kelvins.

A sample table showing coefficients of thermal expansion for selected materials is given below.

Approximate Coefficients of Thermal Expansion at 20°C		
Material	α (10⁻⁶/°C)	β (10⁻⁶/°C)
Aluminum	23	69
Concrete	12	36
Diamond	1	3
Glass	9	27
Stainless Steel	17	51
Water*	69	207

Water actually expands when it freezes, so calculations near the freezing point of water require a more detailed analysis than is provided here.

4.09 Q: A concrete railroad tie has a length 2.45 meters on a hot, sunny, 35°C day. What is the length of the railroad tie in the winter when the temperature dips to -25°C?

4.09 A: First find the change in the tie's length.

$$\Delta l = \alpha l_0 \Delta T = (12 \times 10^{-6} \, /_{°C})(2.45m)(-60°C)$$

$$\Delta l = -0.0018m$$

Then, you can find the railroad tie's final length using the tie's initial length and its change in length.

$$\Delta l = l - l_0 \rightarrow l = l_0 + \Delta l$$

$$l = 2.45m + (-0.0018m) = 2.448m$$

4.10 Q: An aluminum rod has a length of exactly one meter at 300K. How much longer is it when placed in a 400°C oven?

4.10 A: Since the temperatures are given in two different sets of units, you first need to find the total temperature shift in consistent units (for example, Kelvins).

$$T_K = T_{°C} + 273.15 = 400 + 273.15 = 673.15K$$

The shift in temperature, therefore, must be 673.15K-300K, or 373.15K. Next, you can use the equation for linear expansion to find the shift in the rod's length.

$$\Delta l = \alpha l_0 \Delta T = (23 \times 10^{-6} \; /_K)(1m)(373.15K)$$

$$\Delta l = 0.0086m$$

4.11 Q: A glass of water with volume 1 liter is completely filled at 5°C. How much water will spill out of the glass when the temperature is raised to 85°C?

4.11 A: In this situation, both the glass and the water within will expand as the temperature rises. You can treat both the glass and the water as a volume expansion. Start by finding the expansion of the water.

$$\Delta V = \beta V_0 \Delta T = (207 \times 10^{-6} \; /_{°C})(1L)(80°C) = 0.017L$$

In a similar fashion, you can find the expansion of the glass.

$$\Delta V = \beta V_0 \Delta T = (27 \times 10^{-6} \; /_{°C})(1L)(80°C) = 0.002L$$

The amount of water spilling out is equal to the difference between the water's expansion and the glass's expansion, or 0.015 liters.

Heat

Heat is the transfer of thermal energy from one object to another object due to a difference in temperature. Heat always flows from warmer objects to cooler objects. The symbol for heat in physics is Q, with positive values of Q representing heat flowing into an object, and negative values of Q representing heat flowing out of an object.

When heat flows into or out of an object, the amount of temperature change depends on the material. The amount of heat required to change one kilogram of a material by one degree Celsius (or one Kelvin) is known as the material's specific heat (or specific heat capacity), represented by the symbol C.

Specific Heats of Selected Materials	
Material	C (J/kg·K)
Aluminum	897
Concrete	850
Diamond	509
Glass	840
Helium	5193
Water	4181

The relationship between heat and temperature is quantified by the following equation, where Q is the heat transferred, m is the mass of the object, C is the specific heat, and ΔT is the change in temperature (in degrees Celsius or Kelvins).

$$Q = mC\Delta T$$

4.12 Q: A half-carot diamond (0.0001 kg) absorbs five Joules of heat. How much does the temperature of the diamond increase?

4.12 A: $Q = mC\Delta T \rightarrow \Delta T = \dfrac{Q}{mC}$

$\Delta T = \dfrac{5J}{(0.0001kg)(509\,\mathrm{J}/_{kg \bullet K})} = 98K$

4.13 Q: A three-kilogram aluminum pot is filled with five kilograms of water. How much heat is absorbed by the pot and water when both are heated from 25°C to 95°C?

4.13 A: Using the table of specific heats, you can find the heat added to each item separately, and then combine them to get the total heat added.

$Q_{Al} = m_{Al}C_{Al}\Delta T = (3kg)(897\,\mathrm{J}/_{kg \bullet K})(70K) = 1.88 \times 10^5 J$

$Q_{H_2O} = m_{H_2O}C_{H_2O}\Delta T = (5kg)(4181\,\mathrm{J}/_{kg \bullet K})(70K) = 1.46 \times 10^6 J$

The total heat absorbed, therefore, must be 1.65×10⁶ Joules.

4.14 Q: Two solid metal blocks are placed in an insulated container. If there is a net flow of heat between the blocks, they must have different

(A) initial temperatures

(B) melting points

(C) specific heats

(D) heats of fusion

4.14 A: (A) since heat flows from warmer objects to cooler objects.

Heat can be transferred from one object to another by three different methods: conduction, convection, and radiation. **Conduction** is the transfer of heat along an object due to the particles comprising the object colliding. When you stick an iron rod in a fire, the end in the fire warms up, but over time, the particles comprising the iron rod near the fire move more quickly, colliding with other particles in the iron speeding them up, and so on, and so on, resulting in heat transfer down the length of the iron rod until the end you're holding far away from the fire becomes very hot!

Convection, on the other hand, is a result of the energetic (heated) particles moving from one place to another. A great example of this is a convection oven. In a convection oven, air molecules are heated near the burner or electrical element, and then circulated throughout the oven, transferring the heat throughout the entire oven's volume. Convection is typically limited to fluids.

Radiation is the transfer of heat through electromagnetic waves. Think of a campfire or fireplace on a cold evening. When you want to warm up, you place your hands up in front of you, allowing your hands to absorb the maximum amount of electromagnetic waves (mostly infrared) coming from the fire, making you nice and toasty!

Looking more closely at conduction specifically, the rate of heat transfer (H), which is the energy transferred to the system (Q) per unit time (Δt), measured in joules per second, or watts, depends on the magnitude of the temperature difference across the object (ΔT), the cross-sectional area of the object (A), the length of the object (L) and the thermal conductivity of the material (k). Thermal conductivities are typically provided to you in a problem, or you can look them up in a table of thermal conductivities.

$$H = \frac{Q}{\Delta t} = \frac{kA\Delta T}{L}$$

Thermal Conductivities of Selected Materials	
Material	**k (J/s·m·K)**
Aluminum	237
Concrete	1
Copper	386
Glass	0.9
Stainless Steel	16.5
Water	0.6

4.15 Q: Find the rate of heat transfer through a 5 mm thick glass window with a cross-sectional area of 0.4 m² if the inside temperature is 300K and the outside temperature is 250K.

4.15 A: $H = \dfrac{kA\Delta T}{L} = \dfrac{(0.9\,{}^{J}\!/_{s \bullet m \bullet K})(0.4m^2)(300K - 250K)}{0.005m} = 3600W$

4.16 Q: One end of a 1.5-meter-long stainless steel rod is placed in an 850K fire. The cross-sectional radius of the rod is 1 cm, and the cool end of the rod is at 300K. Calculate the rate of heat transfer through the rod.

4.16 A: To solve this problem, you must first find the cross-sectional area of the rod.

$$A = \pi r^2 = \pi (.01m)^2 = 3.14 \times 10^{-4} m^2$$

Next, calculate the heat transfer through the rod.

$$H = \frac{kA\Delta T}{L}$$

$$H = \frac{(16.5\,{}^{J}\!/_{s \bullet m \bullet K})(3.14 \times 10^{-4} m^2)(850K - 300K)}{1.5m} = 1.9W$$

4.17 Q: A glass-walled furnace 20 cm thick sees 1100°C on the furnace side and 200°C on its outside wall. Determine the heat transfer rate per unit area.

4.17 A: Recognizing that the heat transfer rate per unit area is H/A:

$$H = \frac{kA\Delta T}{L} \rightarrow \frac{H}{A} = \frac{k\Delta T}{L} = \frac{(0.9\,J\!/_{s\cdot m\cdot K})(1100°C - 200°C)}{(0.2m)} = 4050\,J\!/_{s\cdot m^2}$$

4.18 Q: A man living in a log cabin steps out of bed in his bare feet onto a stone floor. As he crosses from the stone floor to the wood floor, the wood floor feels warmer to him. This is most likely because

(A) the average kinetic energy of the molecules of the wood floor is larger than the average kinetic energy of the molecules of the stone floor.

(B) the total internal energy of the molecules of the wood floor is greater than the total internal energy of the molecules of the stone floor.

(C) the thermal conductivity of the stone floor is greater than the thermal conductivity of the wood floor and conducts heat out of the man's feet more quickly.

(D) the stone floor is cooler than the wood floor.

4.18 A: (C) the thermal conductivity of the stone floor is greater than the thermal conductivity of the wood floor and conducts heat out of the man's feet more quickly.

4.19 Q: Which method of heat transfer functions in a vacuum?
(A) Conduction
(B) Convection
(C) Radiation
(D) Expansion

4.19 A: (C) Radiation does not require matter to function.

4.20 Q: Convection is most likely to play a significant role in heat transfer in which of the following materials? Select two answers.
(A) An aluminum block
(B) A glass of water
(C) Helium inside a balloon
(D) Wood burning in a fire

4.20 A: (B) and (C). Convection is typically a fluid process.

4.21 Q: A match is held under a paper cup filled with water. The paper does not burn because

(A) the paper cup cannot become significantly warmer than the water it contains.

(B) water has a low thermal conductivity.

(C) paper has a low thermal conductivity.

(D) the paper cup is wet.

4.21 A: (A) the paper cup cannot become significantly warmer than the water it contains. Water boils at 100°C, so the paper cannot become significantly warmer than 100°C while water remains in the cup.

Phase Changes*

As you know, matter can exist in different states. These states include solids, liquids, gases, and plasmas. You're probably familiar with solids, liquids, and gases already. Plasmas are energetic gases that have been ionized so that they can conduct electricity (examples include stars, lightning, neon signs, etc.).

When matter changes state, its internal energy changes, so the kinetic energy of its constituent particles changes. As it is changing from one state to another, the change in energy is reflected in the bonds between the particles, and therefore the temperature of the object doesn't change. Once the state change is complete, however, changes in energy are again observed in the form of changes in temperature.

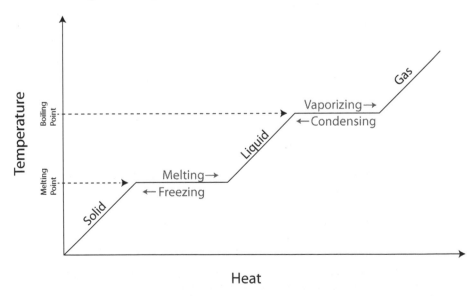

The energy required to change a specific material's state is known as the material's latent heat of transformation (L). When an object transitions from the solid to liquid phase, you use the latent heat of fusion (L_f). When an object transitions from the liquid to the gaseous phase, you use the latent heat of vaporization (L_v). You can calculate the energy required for a material to change phases using the following formula, where Q is the heat added, m is the object's mass, and L is the material's specific latent heat of transformation.

$$Q = mL$$

Latent Heats of Selected Materials				
Material	**L_f (kJ/kg)**	**Melting Point (°C)**	**L_v (kJ/kg)**	**Boiling Point (°C)**
Aluminum	399	659	10,500	2327
Helium	N/A	N/A	21	-269
Hydrogen	58	-259	455	-253
Lead	25	327	871	1750
Water	334	0	2260	100

4.22 Q: The graph below represents a cooling curve for 10 kilograms of a substance as it cools from a vapor at 160°C to a solid at 20°C. Energy is removed from the sample at a constant rate.

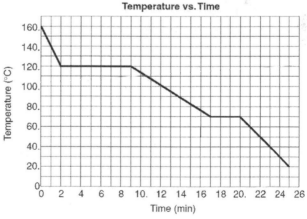

Temperature vs. Time

While the substance is cooling during the liquid phase, the average kinetic energy of the molecules of the substance

(A) decreases

(B) increases

(C) remains the same

4.22 A: (A) decreases since the temperature decreases, and average kinetic energy is related to temperature.

4.23 Q: Based on the graph of the previous problem, the melting point of the substance is

(A) 0°C

(B) 70°C

(C) 100°C

(D) 120°C

4.23 A: (B) 70°C.

4.24 Q: How much heat must be added to a 10 kg lead bar to change the bar from a solid to a liquid at 327°C?

4.24 A: $Q_f = mL_f = (10kg)(25000\,{}^J\!/\!_{kg}) = 250,000J$

4.25 Q: How much heat must be added to 1 kg of water to change it from a 50°C to 100°C steam at standard pressure?

4.25 A: To solve this problem, you must find both the amount of heat required to change the temperature of the water, as well as the amount of heat required to change the state of the water.

$Q = mC\Delta T + mL_v = m(C\Delta T + L_v)$

$Q = 1kg(4181\,{}^J\!/\!_{kg \bullet K} \times 50°C + 2260000\,{}^J\!/\!_{kg}) = 2.47 \times 10^6\,J$

4.26 Q: The graph below shows temperature vs. time for one kilogram of an unknown material as heat is added at a constant rate.

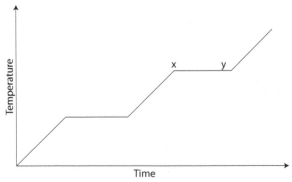

During interval *xy*, the material experiences
(A) a decrease in internal energy and a phase change.
(B) an increase in internal energy and a phase change.
(C) no change in internal energy and a phase change.
(D) no change in internal energy and no phase change.

4.26 A: (B) an increase in internal energy and a phase change.

Ideal Gas Law

In studying the behavior of gases in confined spaces, it is useful to limit ourselves to the study of ideal gases. **Ideal gases** are theoretical models of real gases, which utilize a number of basic assumptions to simplify their study. These assumptions include treating the gas as being comprised of many particles which move randomly in a container. The particles are, on average, far apart from one another, and they do not exert forces upon one another unless they come in contact in an elastic collision.

Under normal conditions such as standard temperature and pressure, most gases behave in a manner quite similar to an ideal gas. Heavy gases as well as gases at very low temperatures or very high pressures are not as well modeled by an ideal gas.

The **Ideal Gas Law** relates the pressure, volume, number of particles, and temperature of an ideal gas in a single equation, and can be written in a number of different ways.

$$PV = nRT = Nk_B T$$

In this equation, P is the pressure of the gas (in Pascals), V is the volume of the gas (in cubic meters), n is the number of moles of gas, N is the number of molecules of gas, R is the universal gas constant equal to 8.31 J/mol·K (which is also 0.08206 L·atm/mol·K), k_B is Boltzmann's Constant (1.38×10^{-23} J/K), and T is the temperature, in Kelvins. To convert from molecules to moles, you can use Avogadro's Number ($N_0 = 6.02 \times 10^{-23}$ molecules/mole):

$$n = \frac{N}{N_0}$$

Note that a Pascal multiplied by a cubic meter is a newton-meter, or Joule. As well, Boltzmann's constant is the ideal gas law constant divided by Avogadro's number.

4.27 Q: How many moles of an ideal gas are equivalent to 3.01×10^{24} molecules?

4.27 A: $n = \dfrac{N}{N_0} = \dfrac{3.01 \times 10^{24}\, molecules}{6.02 \times 10^{23}\, molecules/_{mole}} = 5\, moles$

4.28 Q: Find the number of molecules in 0.4 moles of an ideal gas.

4.28 A: $n = \dfrac{N}{N_0} \rightarrow N = nN_0 \rightarrow$

$N = (0.4 moles)(6.02 \times 10^{23}\, molecules/_{mole}) = 2.41 \times 10^{23}\, molecules$

4.29 Q: How many moles of gas are present in a 0.3 m³ bottle of carbon dioxide held at a temperature of 320K and a pressure of 1×10⁶ Pascals?

4.29 A: $PV = nRT \rightarrow n = \dfrac{PV}{RT} = \dfrac{(1 \times 10^{6}\, Pa)(0.3 m^{3})}{(8.31\, J/_{mol \bullet K})(320K)} = 113\, moles$

4.30 Q: A cubic meter of carbon dioxide gas at room temperature (300K) and atmospheric pressure (101,325 Pa) is compressed into a volume of 0.1 m³ and held at a temperature of 260K. What is the pressure of the compressed carbon dioxide?

4.30 A: Since the number of moles of gas is constant, you can simplify the ideal gas equation into the combined gas law by setting the initial pressure, volume, and temperature relationship equal to the final pressure, volume, and temperature relationship.

$$\frac{P_1 V_1}{T_1} = nR = \frac{P_2 V_2}{T_2}$$

Since you know all the quantities in this equation except for the final pressure, you can solve for the final pressure directly.

$$P_2 = \frac{P_1 V_1 T_2}{T_1 V_2} = \frac{(101,325 Pa)(1 m^{3})(260K)}{(300K)(0.1 m^{3})} = 878,000 Pa$$

4.31 Q: One mole of helium gas is placed inside a balloon. What is the pressure inside the balloon when the balloon rises to a point in the atmosphere where the temperature is -12°C and the volume of the balloon is 0.25 cubic meters?

4.31 A: First you must convert the temperature from degrees Celsius to Kelvins.

$$T_K = T_{°C} + 273.15 = -12°C + 273.15 = 261.15K$$

Next, you can use the ideal gas law to solve for the pressure inside the balloon.

$$PV = nRT \rightarrow P = \frac{nRT}{V} \rightarrow$$

$$P = \frac{(1mole)(8.31\,\text{/}_{mol \bullet K})(261.15K)}{(.25m^3)} = 8680\,Pa$$

4.32 Q: Five sealed containers hold an ideal gas. Rank the temperatures of the ideal gas in the cylinders from highest to lowest based on the diagram below.

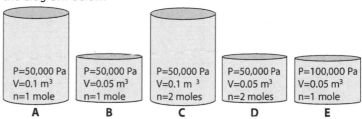

4.32 A: A=E, B=C, D (Apply PV=nRT, so T=PV/nR)

4.33 Q: Four cylinders labeled A through D are filled with an ideal gas and are at the same temperature. In each case, a piston pushes down on the gas with the given force. The cross-sectional area of each piston is given in the diagram. Rank the pressure in each cylinder from highest to lowest given that each setup is in equilibrium.

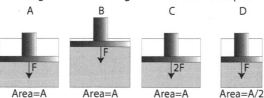

4.33 A: C=D, A=B. Pressure is force divided by area.

It's also quite straightforward to find the total internal energy of an ideal gas. Recall that the average kinetic energy of the particles of an ideal gas are described by the formula:

$$K_{avg} = \frac{3}{2}k_B T$$

The total internal energy of an ideal gas can be found by multiplying the average kinetic energy of the gas's particles by the number of particles (N) in the gas. Therefore, the internal energy of the gas can be calculated using:

$$U = N \times K_{av} \quad \underset{K_{av}=\frac{3}{2}k_B T}{\overset{N=nN_0}{\longrightarrow}} U = \frac{3}{2}nN_0 k_B T \quad \overset{N_0 k_B = R}{\longrightarrow}$$

$$U = \frac{3}{2}nRT$$

It is important to understand that the speeds of the individual atoms and/or molecules in the system can have a wide range of speeds, which are in an ongoing state of flux as momentum is transferred due to the many collisions occurring in the material at the microscopic scale. This is readily depicted by looking at a statistical distribution of particle speeds, modeled by the Maxwell Speed Distribution, shown below for hydrogen, helium, and nitrogen.

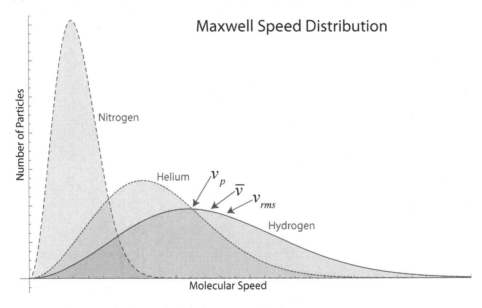

The distribution shows the wide range of particle speeds in an ideal gas. Note that hydrogen, the smallest particle, has the largest speeds and the widest distribution, while nitrogen, the heaviest particle, has smaller speeds and a more narrow distribution. You can also characterize these distributions by looking at the most probable speed (v_p), mean speed ($\bar{v}$), and the root-mean-square speed (v_{rms}), which are shown in the diagram for the hydrogen plot. The root-mean-square speed, or rms speed, is within 10 percent of the mean speed, and is found by squaring all the speeds, averaging the squares, and then taking the square root. It is often times used in place of the mean speed of the particles due to its simplicity of calculation as well as its use in a variety of more detailed calculations.

$$v_{rms} = \sqrt{\frac{3k_B T}{m}} = \sqrt{\frac{3RT}{M}}$$

The rms speed can be found by taking the square root of 3 times Boltzmann's constant (k_B) multiplied by the temperature of the ideal gas (T), divided by the mass of the individual molecule (m). Alternately, you can take the square root of 3 times the universal gas constant (R) multiplied by the temperature of the ideal gas (T), divided by the molar mass (M) of the ideal gas.

4.34 Q: Find the internal energy of 5 moles of oxygen at a temperature of 300K.

4.35 A: $U = \frac{3}{2}nRT = \frac{3}{2}(5moles)(8.31\ /_{mol\bullet K})(300K) = 18.7kJ$

4.35 Q: What is the temperature of 20 moles of argon with a total internal energy of 100 kJ?

4.35 A: $U = \frac{3}{2}nRT \rightarrow T = \dfrac{2U}{3nR} \rightarrow$

$$T = \frac{2(100,000J)}{3(20moles)(8.31\ /_{mol\bullet K})} = 401K$$

4.36 Q: A gas is held in a sealed container. Which of the following changes will increase the pressure? Select two answers.
(A) Doubling the temperature and halving the volume.
(B) Halving the temperature and doubling the volume.
(C) Tripling the temperature and doubling the volume.
(D) Doubling the temperature and tripling the volume.

4.36 A: (A) and (C) will both increase the pressure according to the Ideal Gas Law (P=nRT/V).

4.37 Q: Two gases with the same density and distribution of speeds sit in identical sealed containers. The molecules of gas 2 have twice the mass of the molecules of gas 1. Which of the following statements are true? Select two answers.

(A) Gas 1 has a higher pressure than Gas 2.

(B) Gas 2 has a higher pressure than Gas 1.

(C) Gas 1 has a higher temperature than Gas 2.

(D) Gas 2 has a higher temperature than Gas 1.

4.37 A: (B) and (D). Given the same distribution of speeds, the gas with the bigger mass must have the higher temperature (which you can determine by analyzing the v_{rms} equation or the average kinetic energy equation), so D must be true. If gas 2 has a higher temperature than gas 1, it must also have a higher pressure, which you can determine from PV=nRT.

4.38 Q: Four samples of a gas, labeled A through D, are prepared such that each sample has the same number of molecules but are held at different temperatures. The diagram at right shows the speed distribution of molecules in the samples. Rank the temperatures of the gases from highest to lowest.

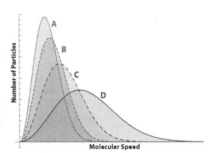

4.38 A: D, C, B, A

Thermodynamics

Thermodynamics, which began as an effort to increase the efficiency of steam engines in the early 1800s, can be thought of as the study of the relationship between heat transferred to or from an object, and the work done on or by an object. Both heat and work deal with the transfer of energy, but heat involves energy transfer due to a temperature difference.

The zeroth law of thermodynamics (don't blame me, I didn't name it!) states that if object A is in thermal equilibrium with object B, and object B is in thermal equilibrium with object C, then objects A and C must be in thermal equilibrium with each other. This law is so intuitive it almost doesn't need stating, but in defining proofs of the 1st and 2nd laws of thermodynamics, scientists realized they needed this law specifically stated to complete their proofs.

The first law of thermodynamics is really a restatement of the law of conservation of energy. Specifically, It states that the change in the internal energy of a closed system is equal to the heat added to the system plus the work done on the system, and is written as:

$$\Delta U = Q + W$$

In this equation it is important to note the sign conventions, where a positive value for heat, Q, represents heat added to the system, and a positive value for work, W, indicates work done on the gas. If energy were being pulled from the system, as in heat taken from the system or work done by the system, those quantities would be negative.

In most cases, you'll utilize the first law of thermodynamics to analyze the behavior of ideal gases, which can be streamlined by analyzing the definition of work on a gas.

$$W = Fd \xrightarrow{F=PA} W = PAd \xrightarrow{\Delta V=-Ad} W = -P\Delta V$$

If work is force multiplied by displacement, and pressure is force over area, force can be replaced with pressure multiplied by area. The area multiplied by the displacement gives you the change in volume of the gas. Due to the sign convention that work done on the gas is positive (corresponding to a decrease in volume), you can write work as W=-PΔV.

4.39 Q: Five thousand joules of heat is added to a closed system, which then does 3000 joules of work. What is the net change in the internal energy of the system?

4.39 A: Keeping in mind the positive/negative sign convention:
$$\Delta U = Q + W = 5000J - 3000J = 2000J$$

4.40 Q: A liquid is changed to a gas at atmospheric pressure (101,325 Pa). The volume of the liquid was 5×10⁻⁶ m³. The volume of gas is 5×10⁻³ m³. How much work was done in the process?

4.40 A: $W = -P\Delta V = -P(V_f - V_i) \rightarrow$
$$W = -(101,325\,Pa)(5\times10^{-3}\,m^3 - 5\times10^{-6}\,m^3) = -506J$$

Pressure-Volume Diagrams (PV diagrams) are useful tools for visualizing the thermodynamic processes of gases. These diagrams show pressure on the y-axis, and volume on the x-axis, and are used to describe the changes undergone by a set amount of gas. Because the amount of gas remains constant, a PV diagram not only tells you pressure and volume, but can also be used to determine the temperature of a gas when combined with the ideal gas law. A Sample PV diagram is shown at right, showing two states of the gas, state A and state B.

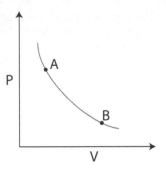

In transitioning from state A to state B, the volume of the gas increases, while the pressure of the gas decreases. In transitioning from state B to state A, the volume of the gas decreases, while the pressure increases. Because the work done on the gas is given by W=-PΔV, you can find the work done on the gas graphically from the PV diagram by taking the area under the curve. Because of the positive/ negative sign convention, as the volume of gas expands the gas does work (W is negative), and as the gas compresses, work is done on the gas (W is positive).

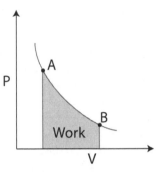

4.41 Q: Using the PV diagram below, find the amount of work required to transition from state A to B, and then the amount of work required to transition from state B to state C.

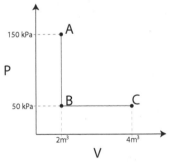

4.41 A: The amount of work in moving from state A to B is equal to the area under the graph for that transition. Since there is no area under the straight line, no work was done. The work in moving from state B to state C can be found by taking the area under the line in the PV diagram.

$$W = -P\Delta V = -P(V_f - V_i) \rightarrow$$

$$W = -(50000\,Pa)(4m^3 - 2m^3) = -1 \times 10^5\,J$$

Note that the work is negative, indicating the gas did work, which correlates with the gas expanding.

In exploring ideal gas state changes, there are a number of state changes in which one of the characteristics of the gas or process remain constant, and are illustrated on the PV diagram below.

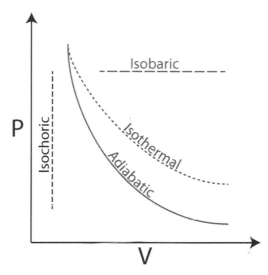

The types of processes include:

- **Adiabatic** — Heat (Q) isn't transferred into or out of the system.
- **Isobaric** — Pressure (P) remains constant.
- **Isochoric** — Volume (V) remains constant.
- **Isothermal** — Temperature (T) remains constant.

In an **adiabatic** process, heat flow (Q) is zero. Applying the first law of thermodynamics, if ΔU=Q+W, and Q is 0, the change in internal energy of the gas must be equal to the work done on the gas (ΔU=W).

In an **isobaric** process, pressure of the gas remains constant. Because pressure is constant, the PV diagram for an isobaric process shows a horizontal line. Further, applying this to the ideal gas law, you find that V/T must remain constant for the process.

In an **isochoric** process, the volume of the gas remains constant. The PV diagram for an isochoric process is a vertical line. Because W=-PΔV, and ΔV=0, the work done on the gas is zero. This is also reflected graphically in the PV diagram. Work can be found by taking the area under the PV graph, but the area under a vertical line is zero. Applying this to the ideal gas law, you find that P/T must remain constant for an isochoric process.

In an **isothermal** process, temperature of the gas remains constant. Lines on a PV diagram describing any process held at constant temperature are therefore called isotherms. In an isothermal process, the product of the pressure and the volume of the gas remains constant. Further, because temperature is constant, the internal energy of the gas must remain constant, therefore Q=-W.

4.42 Q: An ideal gas undergoes an adiabatic expansion, doing 2000 joules of work. How much does the gas's internal energy change?

4.42 A: Since the process is adiabatic, Q=0, therefore:

$$\Delta U = Q + W \xrightarrow{\;Q=0\;} \Delta U = W = -2000J$$

4.43 Q: An ideal gas is represented in four distinct states in the PV diagram at right. Rank the temperatures of the ideal gas in the states from highest temperature to lowest temperature.

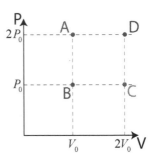

4.43 A: D, A=C, B

4.44 Q: Heat is removed from an ideal gas as its pressure drops from 200 kPa to 100 kPa. The gas then expands from a volume of 0.05 m³ to 0.1 m³ as shown in the PV diagram below. If curve AC represents an isotherm, find the work done by the gas and the heat added to the gas.

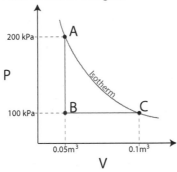

4.44 A: The work done by the gas in moving from A to B is zero, as the area under the graph is zero. In moving from B to C, however, the work done by the gas can be found by taking the area under the graph.

$$W = -P\Delta V = -(100,000\,Pa)(0.1m^3 - 0.05m^3) = -5000J$$

The negative sign indicates that 5000 joules of work was done by the gas. Since AC is on an isotherm, the temperature of the gas must remain constant. Therefore, the gas's internal energy must remain constant. Knowing that ΔU=Q+W, if ΔU=0, then Q must be equal to -W; therefore 5000 joules must have been added to the gas.

4.45 Q: One mole of an ideal gas undergoes a series of changes in pressure and volume as it follows path ABCA on the PV diagram. Using the PV diagram, find the temperature of the gas at each point, then fill in the information table by finding the change in internal energy, work done on, and heat added to the gas for each process.

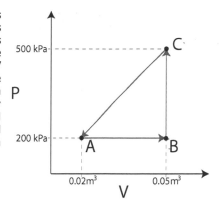

Process	ΔU	Q	W
A to B	~	+	-
B to C	+	-	◯
C to A	+	-	

4.45 A: Start by finding the temperature at each point using PV=nRT.

$$PV = nRT \rightarrow T = \frac{PV}{nR}$$

$$T_A = \frac{P_A V_A}{nR} = \frac{(200,000\,Pa)(0.02m^3)}{1 \times 8.31\,{}^J\!/_{mol \bullet K}} = 481K$$

$$T_B = \frac{P_B V_B}{nR} = \frac{(200,000\,Pa)(0.05m^3)}{1 \times 8.31\,{}^J\!/_{mol \bullet K}} = 1203K$$

$$T_C = \frac{P_C V_C}{nR} = \frac{(500,000\,Pa)(0.05m^3)}{1 \times 8.31\,{}^J\!/_{mol \bullet K}} = 3008K$$

Next, fill in the table for isobaric process A to B.

$$W = -P\Delta V = -(200,000\,Pa)(0.05m^3 - 0.02m^3) = -6000J$$

$$\Delta U = \tfrac{3}{2}nR\Delta T = \tfrac{3}{2}nR(T_B - T_A) \rightarrow$$

$$\Delta U = \tfrac{3}{2}(1)(8.31\,{}^J\!/_{mol \bullet K})(1203K - 481K) = 9000J$$

$$\Delta U = Q + W \rightarrow Q = \Delta U - W \rightarrow$$

$$Q = 9000J - (-6000J) = 15000J$$

Now you can determine the work done in moving from B to C is zero because it is an isochoric (constant volume) process. Then, fill in the rest of the table for process B to C.

$$\Delta U = \tfrac{3}{2}nR\Delta T = \tfrac{3}{2}nR(T_C - T_B) \rightarrow$$
$$\Delta U = \tfrac{3}{2}(1)(8.31\,\tfrac{J}{mol \bullet K})(3008K - 1203K) = 22{,}500\,J$$

$$\Delta U = Q + W \rightarrow Q = \Delta U - W \rightarrow$$
$$Q = 22{,}500\,J - 0 = 22{,}500\,J$$

Finally, you can determine the work done in moving from C to A by taking the area under the graph (noting that work is done on the gas as it is compressed, therefore it must be positive), then completing the remaining calculations in a similar fashion.

$$W = Area = \tfrac{1}{2}bh + lw = \tfrac{1}{2}(0.03m^3)(300000Pa) + 6000\,J$$
$$W = 10{,}500\,J$$

$$\Delta U = \tfrac{3}{2}nR\Delta T = \tfrac{3}{2}nR(T_A - T_C) \rightarrow$$
$$\Delta U = \tfrac{3}{2}(1)(8.31\,\tfrac{J}{mol \bullet K})(481K - 3008K) = -31{,}500\,J$$

$$\Delta U = Q + W \rightarrow Q = \Delta U - W \rightarrow$$
$$Q = -31{,}500\,J - 10{,}500\,J = -42{,}000\,J$$

Process Summary Table			
Process	ΔU	Q	W
A to B	9,000	15,000	-6,000
B to C	22,500	22,500	0
C to A	-31,500	-42,000	10,500

4.46 Q: An ideal gas can move from State 1 to State 2 via three different possible paths: Path A, Path B, and Path C. Which of the following statements are true? Select two answers.

(A) Change in temperature is the same for all three paths.

(B) Heat absorbed is the same for all three paths.

(C) Work done by the gas is the same for all three paths.

(D) Change in internal energy is the same for all three paths.

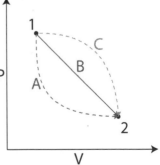

4.46 A: (A) and (D) are true, as both the change in temperature and the change in internal energy are path independent.

4.47 Q: Four processes compose a thermodynamic cycle as shown in the PV diagram for the ideal gas. Which point represents the highest temperature of the gas, A, B, C, or D? Which process represents the most work done on the gas, AB, BC, CD, or DA?

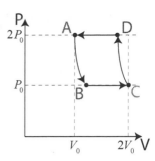

4.47 A: Highest temperature: D (highest PV combination)

Most work: DA (work is area under the curve)

4.48 Q: A fixed amount of ideal gas is subjected to one of two process cycles, labeled ABCD and EFGH, depicted on the PV diagram below. Use the diagram to answer the following questions.

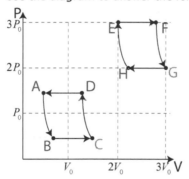

A) Compare the work done by the gas for process BC to the work done by the gas for process EF.

B) Compare the work done on the gas for process DA to the work done on the gas for process EF.

C) Compare the work done by the gas for process ABCDA to the work done by the gas for process EFGHE.

4.48 A: A) The work done by the gas is the area under the line BC compared to the area under the line EF; therefore the work done by the gas is greater for process EF.

B) The work done on the gas is the area under the line DA, which is positive since the gas is being compressed. The work done on the gas is the opposite of the area under the line EF; therefore the work done on the gas for EF is negative. The work done on the gas, therefore, is greater for process DA than process EF.

C) The magnitude of the work done is the same for both cycles since they contain the same area, though the system does work in cycle EFGHE, and work is done on the system in ABCDA. The work done by the gas is therefore greater for process EFGHE.

NOTE: When looking at PV diagrams, clockwise processes indicate net work is done by the gas (an engine), while counter-clockwise processes indicate net work is done on the gas.

The second law of thermodynamics can be stated in a variety of ways. One statement of this law says that heat flows naturally from a warmer object to a colder object, and cannot flow from a colder object to a warmer object without an external force doing work on the system. This can be observed quite easily in everyday circumstances. For example, your cold spoon contacting your hot soup never results in your soup becoming hotter and your spoon becoming colder.

The second law of thermodynamics also limits the efficiency of any heat engine, and proves that it is not possible to make a 100 percent efficient heat engine, even if friction were completely eliminated.

Another statement of this law says that the level of entropy, or disorder, in a closed system can only increase or remain the same. This means that your desk will never naturally become more organized without doing work. It also means that you can't drop a handful of plastic building blocks and observe them spontaneously land in an impressive model of a medieval castle. Unfortunately, it even means that no matter how many times Humpty Dumpty falls off his wall, all his pieces on the ground will never end up more organized after he hits the ground compared to before his balance failed him.

Absolute Zero and Nernst's Theorem

Taking another look at the ideal gas law, it is fairly straightforward to show that there exists a direct relationship between temperature and volume of a gas held at a constant pressure. In creating a plot of these characteristics, you find that as you go down in temperature, gases eventually liquify, making it impossible to complete the entire graph. By extrapolating the data down to the point at which the gas would take up zero volume, you find that all gases reach the same point, -273.15°C, also known as absolute zero. This is the zero point of the Kelvin temperature scale.

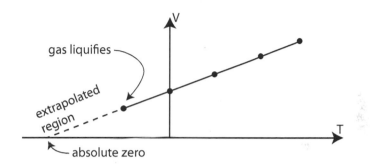

The third and final law of thermodynamics, also known as Nernst's Theorem after its discoverer, Walter Nernst, states that no material can ever be cooled to absolute zero (although materials can get awfully close!).

Test Your Understanding

1. In your own words, state the three laws of thermodynamics.

2. Explain the difference between internal energy, temperature, and heat.

3. If you want your coffee at the hottest possible temperature when you drink it, should you add your cold creamer to the coffee as soon as you receive it, knowing it will be a few minutes before you have a chance to drink the beverage, or keep the creamer and coffee separate until right before you drink the coffee? Explain.

4. Design an experiment to determine the thermal conductivity of a soup can.

5. Find examples of everyday heating systems that function primarily on the basis of conduction, convection, and radiation.

6. In your own words, explain what is meant by the statement "the total entropy of the universe is always increasing." Does this statement violate the law of conservation of energy? Why or why not?

7. Occasionally sick patients with a fever are cooled by placing a damp cloth on their forehead. How does the damp cloth assist in cooling the patient?

8. Why does gas escaping from a pressurized container feel cool?

9. Certain amusement parks now offer cooling machines which spray a fine mist of water for patrons to walk through on hot days. Explain how walking through the mist of water cools people down.

Chapter 5: Electrostatics

"Electricity can be dangerous. My nephew tried to stick a penny into a plug. Whoever said a penny doesn't go far didn't see him shoot across that floor.

I told him he was grounded."

— Tim Allen

Objectives

1. Identify and describe types of electric charge and their origins.
2. Make claims and predictions about natural phenomena utilizing the law of conservation of charge.
3. Describe the differences between conductors and insulators in terms of material characteristics and electric behavior.
4. Use Coulomb's Law to solve problems related to electrical force.
5. Explain how contact forces such as tension, friction, normal, spring, and buoyant forces arise from interatomic electric forces.
6. Compare and contrast Newton's Law of Universal Gravitation with Coulomb's Law.
7. Define, measure, calculate and describe electric field and electric potential difference in various configurations and representations.
8. Solve problems related to charge, electric field, electric potential, and forces.
9. Solve problems involving charged parallel plates and dielectrics.

Electricity and magnetism play a profound role in almost all aspects of life. From the moment you wake up, to the moment you go to sleep (and even while you're sleeping), applications of electricity and magnetism provide tools, light, warmth, transportation, communication, and even entertainment. Despite its widespread use, however, there is much about these phenomena that is not well understood.

Electric Charges

Matter is made up of atoms. Once thought to be the smallest building blocks of matter, scientists now know that atoms can be broken up into even smaller pieces, known as protons, electrons, and neutrons. Each atom consists of a dense core of positively charged protons and uncharged (neutral) neutrons. This core is known as the nucleus. It is surrounded by a "cloud" of much smaller, negatively charged electrons. These electrons orbit the nucleus in distinct energy levels. To move to a higher energy level, an electron must absorb energy. When an electron falls to a lower energy level, it gives off energy.

Most atoms are neutral -- that is, they have an equal number of positive and negative charges, giving a net charge of 0. In certain situations, however, an atom may gain or lose electrons. In these situations, the atom as a whole is no longer neutral and is called an **ion**. If an atom loses one or more electrons, it has a net positive charge and is known as a positive ion. If, instead, an atom gains one or more electrons, it has a net negative charge and is therefore called a negative ion. Like charges repel each other, while opposite charges attract each other. In physics, the charge on an object is represented with the symbol q.

Charge is a fundamental measurement in physics, much as length, time, and mass are fundamental measurements. The fundamental unit of charge is the **coulomb** [C], which is a very large amount of charge. Compare that to the magnitude of charge on a single proton or electron, known as an elementary charge (e), which is equal to 1.6×10^{-19} coulomb. It would take 6.25×10^{18} elementary charges to make up a single coulomb of charge!

5.01 Q: An object possessing an excess of 6.0×10^6 electrons has what net charge?

5.01 A: $6 \times 10^6 \text{ electrons} \bullet \dfrac{-1.6 \times 10^{-19} C}{1 \text{ electron}} = -9.6 \times 10^{-13} C$

5.02 Q: An alpha particle consists of two protons and two neutrons. What is the charge of an alpha particle?

(A) 1.25×10^{19} C

(B) 2.00 C

(C) 6.40×10^{-19} C

(D) 3.20×10^{-19} C

5.02 A: (D) The net charge on the alpha particle is +2 elementary charges.

$$2e \bullet \frac{1.6 \times 10^{-19} C}{1e} = 3.2 \times 10^{-19} C$$

5.03 Q: If an object has a net negative charge of 4 coulombs, the object possesses

(A) 6.3×10^{18} more electrons than protons

(B) 2.5×10^{19} more electrons than protons

(C) 6.3×10^{18} more protons than electrons

(D) 2.5×10^{19} more protons than electrons

5.03 A: (B) $-4C \bullet \dfrac{1e}{1.6 \times 10^{-19} C} = -2.5 \times 10^{19} e$

5.04 Q: Which quantity of excess electric charge could be found on an object?

(A) 6.25×10^{-19} C

(B) 4.80×10^{-19} C

(C) 6.25 elementary charges

(D) 1.60 elementary charges

5.04 A: (B) all other choices require fractions of an elementary charge, while choice (B) is an integer multiple (3e) of elementary charges.

5.05 Q: What is the net electrical charge on a magnesium ion that is formed when a neutral magnesium atom loses two electrons?

(A) -3.2×10^{-19} C

(B) -1.6×10^{-19} C

(C) $+1.6 \times 10^{-19}$ C

(D) $+3.2 \times 10^{-19}$ C

5.05 A: (D) the net charge must be +2e, or $2(1.6 \times 10^{-19}$ C)=3.2×10^{-19} C.

The Standard Model

Where does electric charge come from? To answer that question, you need to dive into the **Standard Model of Particle Physics**. As you've learned previously, the atom is the smallest part of an element (such as oxygen) that has the characteristics of the element. Atoms are made up of very small negatively charged electrons surrounding the much larger nucleus. The nucleus is composed of positively charged protons and neutral neutrons, together known as **nucleons**, as they are found in the nucleus of the atom. The positively charged protons exert a repelling electrical force upon each other, but the strong nuclear force holds the protons and neutrons together in the nucleus.

This completely summarized scientists' understanding of atomic structure until the 1930s, when scientists began to discover evidence that there was more to the picture and that protons and neutrons were made up of even smaller particles. This launched the particle physics movement, which, to this day, continues to challenge the understanding of the entire universe by exploring the structure of the atom.

In addition to standard matter, researchers have discovered the existence of antimatter. **Antimatter** is matter made up of particles with the same mass as regular matter particles, but opposite charges and other characteristics. An **antiproton** is a particle with the same mass as a proton, but a negative (opposite) charge. A **positron** has the same mass as an electron, but a positive charge. An **antineutron** has the same mass as a neutron, but has other characteristics opposite that of the neutron.

When a matter particle and its corresponding antimatter particle meet, the particles may combine to **annihilate** each other, resulting in the complete conversion of both particles into energy.

This book has dealt with many types of forces, ranging from contact forces such as tensions and normal forces to field forces such as the electrical force and gravitational force. When observed from their most basic aspects, however, all observed forces in the universe can be consolidated into four fundamental forces. They are, from strongest to weakest:

1. **Strong Nuclear Force**: holds protons and neutrons together in the nucleus
2. **Electromagnetic Force**: electrical and magnetic attraction and repulsion
3. **Weak Nuclear Force**: responsible for radioactive beta decay
4. **Gravitational Force**: attractive force between objects with mass

Understanding these forces remains a topic of scientific research, with current work exploring the possibility that forces are actually conveyed by an exchange of force-carrying particles.

5.06 Q: The particles in a nucleus are held together primarily by the
(A) strong force
(B) gravitational force
(C) electrostatic force
(D) magnetic force

5.06 A: (A) the strong nuclear force holds protons and neutrons together in the nucleus.

5.07 Q: Which statement is true of the strong nuclear force?
(A) It acts over very great distances.
(B) It holds protons and neutrons together.
(C) It is much weaker than gravitational forces.
(D) It repels neutral charges.

5.07 A: (B) The strong nuclear force holds protons and neutrons together.

5.08 Q: The strong force is the force of
(A) repulsion between protons
(B) attraction between protons and electrons
(C) repulsion between nucleons
(D) attraction between nucleons

5.08 A: (D) attraction between nucleons (nucleons are particles in the nucleus such as protons and neutrons).

The current model of sub-atomic structure used to understand matter is known as the Standard Model. Development of this model began in the late 1960s, and has continued through today with contributions from many scientists across the world. The Standard Model explains the interactions of the strong (nuclear), electromagnetic, and weak forces, but has yet to account for the gravitational force.

Although the Standard Model itself is a very complicated theory, the basic structure of the model is fairly straightforward. According to the model, all matter is divided into two categories, known as **hadrons** and the much smaller **leptons**. All of the fundamental forces act on hadrons, which include particles such as protons and neutrons. In contrast, the strong nuclear force doesn't act on leptons, so only three fundamental forces act on leptons, such as electrons, positrons, muons, tau particles and neutrinos.

Hadrons are further divided into **baryons** and **mesons**. Baryons such as protons and neutrons are composed of three smaller particles known as **quarks**. Charges of baryons are always whole numbers. Mesons are composed of a quark and an anti-quark (for example, an up quark and an anti-down quark).

Classification of Matter

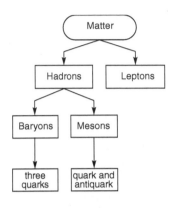

Particles of the Standard Model

Quarks

	up	charm	top
Name			
Symbol	u	c	t
Charge	$+\frac{2}{3}e$	$+\frac{2}{3}e$	$+\frac{2}{3}e$

down	strange	bottom
d	s	b
$-\frac{1}{3}e$	$-\frac{1}{3}e$	$-\frac{1}{3}e$

Leptons

electron	muon	tau
e	μ	τ
−1e	−1e	−1e

electron neutrino	muon neutrino	tau neutrino
ν_e	ν_μ	ν_τ
0	0	0

Note: For each particle, there is a corresponding antiparticle with a charge opposite that of its associated particle.

Scientists have identified six types of quarks. For each of the six types of quarks, there also exists a corresponding anti-quark with an opposite charge. The quarks have rather interesting names: up quark, down quark, charm quark, strange quark, top quark, and bottom quark. Charges on each quark are either one third of an elementary charge, or two thirds of an elementary charge, positive or negative, and the quarks are symbolized by the first letter of their name. For the associated anti-quark, the symbol is the first letter of the anti-quark's name, with a line over the name. For example, the symbol for the up quark is u. The symbol for the anti-up quark is ū.

Similarly, scientists have identified six types of leptons: the electron, the muon, the tau particle, and the electron neutrino, muon neutrino, and tau neutrino. Again, for each of these leptons there also exists an associated anti-lepton. The most familiar lepton, the electron, has a charge of -1e. Its anti-particle, the positron, has a charge of +1e.

Since a proton is made up of three quarks, and has a positive charge, the sum of the charges on its constituent quarks must be equal to one elementary charge. A proton is actually comprised of two up quarks and one down quark. You can verify this by adding up the charges of the proton's constituent quarks (uud).

$$+\frac{2}{3}e \ + \ +\frac{2}{3}e \ + \ -\frac{1}{3}e \ = +1e$$

5.09 Q: A neutron is composed of up and down quarks. How many of each type of quark are needed to make a neutron?

5.09 A: The charge on the neutron must sum to zero, and the neutron is a baryon, so it is made up of three quarks. To achieve a total charge of zero, the neutron must be made up of one up quark (+2/3e) and two down quarks (-1/3e).

If the charge on a quark (such as the up quark) is (+2/3)e, the charge of the anti-quark (ū) is (-2/3)e. The anti-quark is the same type of particle, with the same mass, but with the opposite charge.

5.10 Q: What is the charge of the down anti-quark?

5.10 A: The down quark's charge is -1/3e, so the anti-down quark's charge must be +1/3e.

5.11 Q: Compared to the mass and charge of a proton, an antiproton has
(A) the same mass and the same charge
(B) greater mass and the same charge
(C) the same mass and the opposite charge
(D) greater mass and the opposite charge

5.11 A: (C) the same mass and the opposite charge.

Conductors and Insulators

Certain materials allow electric charges to move freely. These are called **conductors**. Examples of good conductors include metals such as gold, copper, silver, and aluminum. In contrast, materials in which electric charges cannot move freely are known as **insulators**. Good insulators include materials such as glass, plastic, and rubber. Metals are better conductors of electricity compared to insulators because metals contain more free electrons.

Conductors and insulators are characterized by their resistivity, or ability to resist movement of charge. Materials with high resistivities are good insulators. Materials with low resistivities are good conductors.

Semiconductors are materials which, in pure form, are good insulators. However, by adding small amounts of impurities known as dopants, their resistivities can be lowered significantly until they become good conductors. –

Charging by Conduction

Insulators can be charged by contact, or **conduction**. If you take a balloon and rub it against your hair, some of the electrons from the atoms in your hair are transferred to the balloon. The balloon now has extra electrons, and therefore has a net negative charge. Your hair has a deficiency of electrons, so therefore it now has a net positive charge.

Much like momentum and energy, charge is also conserved in a closed system. Continuing the hair and balloon example, the magnitude of the net positive charge on your hair is equal to the magnitude of the net negative charge on the balloon. The total charge of the hair/balloon system remains zero (neutral). For every extra electron (negative charge) on the balloon, there is a corresponding missing electron (positive charge) in your hair. This known as the law of conservation of charge.

Conductors can also be charged by conduction. If a charged conductor is brought into contact with an identical neutral conductor, the net charge will be shared across the two conductors.

5.12 Q: If a conductor carrying a net charge of 8 elementary charges is brought into contact with an identical conductor with no net charge, what will be the charge on each conductor after they are separated?

5.12 A: Each conductor will have a charge of 4 elementary charges.

5.13 Q: What is the net charge (in coulombs) on each conductor after they are separated?

5.13 A: $q=4e=4(1.6\times10^{-19}$ C$)=6.4\times10^{-19}$ C

5.14 Q: Metal sphere A has a charge of +6 units and an identical metal sphere, B, has a charge of −4 units. If the spheres are brought into contact with each other and then separated, the charge on sphere B will be

(A) 0 units

(B) -1 unit

(C) +1 unit

(D) +5 units

5.14 A: (C) +1 unit.

5.15 Q: Compared to insulators, metals are better conductors of electricity because metals contain more free

(A) protons

(B) electrons

(C) positive ions

(D) negative ions

5.15 A: (B) electrons.

A simple tool used to detect small electric charges known as an **electroscope** functions on the basis of conduction. The electroscope consists of a conducting rod attached to two thin conducting leaves at one end and isolated from surrounding charges by an insulating stopper placed in a flask. If a charged object is placed in contact with the conducting rod, part of the charge is transferred to the rod. Because the rod and leaves form a conducting path and like charges repel each other, the charges are distributed equally along the entire rod and leaf apparatus. The leaves, having like charges, repel each other, with larger charges providing greater leaf separation!

5.16 Q: Separation of the leaves of an electroscope when an object is touched to the electroscope's conducting rod could indicate (choose all that apply):

(A) the object is neutral

(B) the object is negatively charged

(C) the object is positively charged

(D) the object is an insulator

5.16 A: (B) & (C) Separation of the leaves of an electroscope indicate the object touching the electroscope is charge. Whether the charge is positive or negative cannot be determined with just this test.

5.17 Q: The diagram below shows three identical metal spheres, X, Y, and Z, on insulating stands. Spheres X and Z are initially uncharged, while sphere Y is charged to -4 C as shown below.

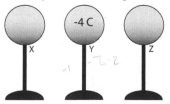

Sphere Y and Sphere Z are brought into contact and then separated. Sphere X and Sphere Y are then brought into contact and separated. What is the final charge on Sphere Y?

5.17 A: -1 C. When Sphere Y and Sphere Z are brought into contact, the charge between the two spheres is charged, leaving -2C on each. When Sphere Y and Sphere X are then placed into contact, the charge is shared again, leaving -1 C on Spheres X and Y.

Charging by Induction

Conductors can also be charged without coming into contact with another charged object in a process known as charging by **induction**. This is accomplished by placing the conductor near a charged object and grounding the conductor. To understand charging by induction, you must first realize that when an object is connected to the earth by a conducting path, known as grounding, the earth acts like an infinite source for providing or accepting excess electrons.

To charge a conductor by induction, you first bring it close to another charged object. When the conductor is close to the charged object, any free electrons on the conductor will move toward the charged object if the object is positively charged (since opposite charges attract) or away from the charged object if the object is negatively charged (since like charges repel).

If the conductor is then "grounded" by means of a conducting path to the Earth, the excess charge is compensated for by means of electron transfer to or from Earth. Then the ground connection is severed. When the originally charged object is moved far away from the conductor, the charges in the conductor redistribute, leaving a net charge on the conductor as shown.

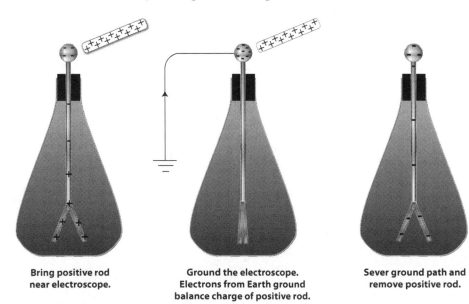

Bring positive rod near electroscope.

Ground the electroscope. Electrons from Earth ground balance charge of positive rod.

Sever ground path and remove positive rod.

Chapter 5: Electrostatics

Charging by Polarization

You can also induce a charge in a charged region in a neutral object by bringing a strong positive or negative charge close to that object. In such cases, the electrons in the neutral object tend to move toward a strong positive charge, or away from a large negative charge. Though the object itself remains neutral, portions of the object are more positive or negative than other parts. In this way, you can attract a neutral object by bringing a charged object close to it, positive or negative. Put another way, a positively charged object can be attracted to both a negatively charged object and a neutral object, and a negatively charged object can be attracted to both a positively charged object and a neutral object.

For this reason, the only way to tell if an object is charged is by repulsion. A positively charge object can only be repelled by another positive charge and a negatively charged object can only be repelled by another negative charge.

5.18 Q: A positively charged glass rod attracts object X. The net charge of object X

(A) may be zero or negative

(B) may be zero or positive

(C) must be negative

(D) must be positive

5.18 A: (A) a positively charged rod can attract a neutral object or a negatively charged object.

5.19 Q: The diagram below shows three neutral metal spheres, x, y, and z, in contact and on insulating stands.

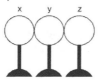

Which diagram best represents the charge distribution on the spheres when a positively charged rod is brought near sphere x, but does not touch it?

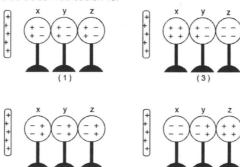

5.19 A: (4) is the correct answer.

5.20 Q: A positively charged metal sphere is lowered by an insulating thread into a grounded metal cup without touching the sides of the cup. The grounding wire is then detached from the metal cup, and the sphere is then removed. Which statement best describes the charge on the cup?

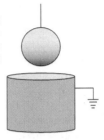

(A) The metal cup is neutral

(B) The metal cup is negatively charged

(C) The metal cup is positively charged

5.20 A: (B) The metal cup is negatively charged. When the positively charged sphere is moved into the cup, negative charges (electrons) are attracted through the ground wire into the cup. The grounding wire is then removed, leaving an excess of electrons on the metal cup. Removing the metal sphere from the cup has no effect on the net charge of the cup so long as the two objects don't come into contact.

Coulomb's Law

Like charges repel and opposite charges attract. In order for charges to repel or attract, they apply a force upon each either, known as the **electrostatic force**. Similar to the manner in which the force of attraction between two masses is determined by the amount of mass and the distance between the masses, as described by Newton's Law of Universal Gravitation, the force of attraction or repulsion is determined by the amount of charge and the distance between the charges.

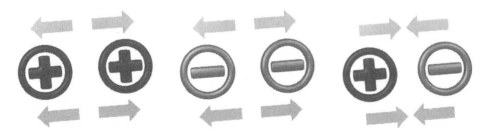

The magnitude of the electrostatic force is described by **Coulomb's Law**, which states that the magnitude of the electrostatic force (F_e) between two objects is equal to a constant, k, multiplied by each of the two charges, q_1 and q_2, and divided by the square of the distance between the charges (r^2). The constant k is known as the **electrostatic constant** and is given as k=8.99×10⁹ N·m²/C² (oftentimes rounded to 9×10⁹ N·m²/C² for simplicity). For mathematical purposes, you may sometimes see k written as $1/4\pi\varepsilon_0$, where ε_0, the permittivity of free space, is 8.85×10⁻¹² C²/N·m².

$$\left|\vec{F}_e\right| = \frac{kq_1q_2}{r^2} = \frac{1}{4\pi\varepsilon_0}\frac{q_1q_2}{r^2}$$

Notice how similar this formula is to the formula for the gravitational force! Both Newton's Law of Universal Gravitation and Coulomb's Law follow the inverse-square relationship, a pattern that repeats many times over in physics. The further you get from the charges, the weaker the electrostatic force. If you were to double the distance from a charge, you would quarter the electrostatic force on a charge.

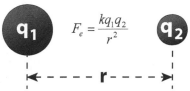

Formally, a positive value for the electrostatic force indicates that the force is a repelling force, while a negative value for the electrostatic force indicates that the force is an attractive force. Because force is a vector, you must assign a direction to it. To determine the direction of the force vector, once you have calculated its magnitude, use common sense to tell you the direction on each charged object. If the objects have opposite charges, they attract each other, and if they have like charges, they repel each other.

5.21 Q: Three protons are separated from a single electron by a distance of 1×10⁻⁶ m. Find the electrostatic force between them. Is this force attractive or repulsive?

5.21 A: $q_1 = 3 \text{ protons} = 3(1.6\times10^{-19}C) = 4.8\times10^{-19}C$

$q_2 = 1 \text{ electron} = 1(-1.6\times10^{-19}C) = -1.6\times10^{-19}C$

$$F_e = \frac{kq_1q_2}{r^2} = \frac{(8.99\times10^9 \frac{N\cdot m^2}{C^2})(4.8\times10^{-19}C)(-1.6\times10^{-19}C)}{(1\times10^{-6}m)^2}$$

$F_e = -6.9\times10^{-16}N \text{ attractive}$

5.22 Q: A distance of 1.0 meter separates the centers of two small charged spheres. The spheres exert gravitational force F_g and electrostatic force F_e on each other. If the distance between the spheres' centers is increased to 3.0 meters, the gravitational force and electrostatic force, respectively, may be represented as

(A) $F_g/9$ and $F_e/9$
(B) $F_g/3$ and $F_e/3$
(C) $3F_g$ and $3F_e$
(D) $9F_g$ and $9F_e$

5.22 A: (A) due to the inverse square law relationships.

5.23 Q: A beam of electrons is directed into the electric field between two oppositely charged parallel plates, as shown in the diagram below.

Electron beam

The electrostatic force exerted on the electrons by the electric field is directed

(A) into the page

(B) out of the page

(C) toward the bottom of the page

(D) toward the top of the page

5.23 A: (D) toward the top of the page because the electron beam is negative, and will be attracted by the positively charged upper plate and repelled by the negatively charged lower plate.

5.24 Q: The centers of two small charged particles are separated by a distance of 1.2×10^{-4} meter. The charges on the particles are $+8.0 \times 10^{-19}$ coulomb and $+4.8 \times 10^{-19}$ coulomb, respectively.

A) Calculate the magnitude of the electrostatic force between these two particles.

B) Sketch a graph showing the relationship between the magnitude of the electrostatic force between the two charged particles and the distance between the centers of the particles.

5.24 A: A) $F_e = \dfrac{kq_1 q_2}{r^2} = \dfrac{(8.99 \times 10^9 \, \frac{N \cdot m^2}{C^2})(8.0 \times 10^{-19} C)(4.8 \times 10^{-19} C)}{(1.2 \times 10^{-4} m)^2}$

$F_e = 2.4 \times 10^{-19} N$

B)

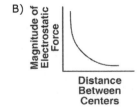

Distance Between Centers

5.25 Q: The diagram below shows a beam of electrons fired through the region between two oppositely charged parallel plates in a cathode ray tube.

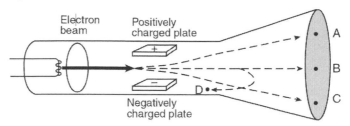

After passing between the charged plates, the electrons will most likely travel path A, B, C, or D?

5.25 A: (A) Electrons are attracted toward positive top plate and repelled by negative bottom plate.

It's important to keep in mind that force is a vector, and therefore has both magnitude and direction, which allows you to apply Coulomb's Law to problems of multiple dimensions. In cases where you have forces due to multiple charges, determine the force due to each individual charge and then add the forces using the superposition principle.

5.26 Q: Three point charges are located at the corners of a right triangle as shown, where $q_1 = q_2 = 3$ μC and $q_3 = -4$ μC. If q_1 and q_2 are each 1 cm from q_3, find the net force on q_3.

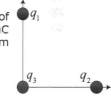

5.26 A: First find the magnitude of the force of charge q_1 on q_3 using Coulomb's Law:

$$\left|\vec{F}_{1,3}\right| = \frac{kq_1q_3}{r^2} = \frac{(8.99\times10^9 \; ^{N\cdot m^2}/_{C^2})(3\times10^{-6}C)(4\times10^{-6}C)}{(.01m)^2} = 1080N$$

Because the charges are of opposite sign, the force is attractive; therefore the force on q_3 is upward. Writing the force of q_1 on q_3 in vector notation:

$$\vec{F}_{1,3} = <0,1080N>$$

Next, find the magnitude of the force of charge q_2 on q_3:

$$\left|\vec{F}_{2,3}\right| = \frac{kq_2q_3}{r^2} = \frac{(8.99\times10^9 \; ^{N\cdot m^2}/_{C^2})(3\times10^{-6}C)(4\times10^{-6}C)}{(.01m)^2} = 1080N$$

Again, because the charges are of opposite sign, the force is attractive; therefore the force on q_3 is to the right. Writing the force of q_2 on q_3 in vector notation:

$$\vec{F}_{2,3} = <1080N,0>$$

You can then add your two force vectors to get the total force acting on charge q_3.

$$\vec{F}_{total_{q3}} = \vec{F}_{1,3} + \vec{F}_{2,3} = <0,1080N> + <1080N,0> = <1080N,1080N>$$

Find the magnitude of the total force using the Pythagorean Theorem and the direction using trigonometry (or common sense as the vertical and horizontal components of the total force are equal in magnitude).

$$\left|\vec{F}_{total_{q3}}\right| = \sqrt{(1080N)^2 + (1080N)^2} = 1527N \, @ \, 45° \text{ North of East}$$

5.27 Q: Two identical charged balls of mass 5 g are hung from the ceiling by a light string of length 20 cm. The total angle between them is 12 degrees. Find the magnitude of the charge on each ball.

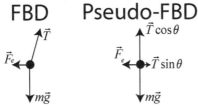

5.27 A: First draw a free body diagram for the charge on a single ball. Noting that the tension is at an angle, you can break it up into horizontal and vertical components to create a pseudo-free body diagram.

FBD   Pseudo-FBD

Next, utilize Newton's 2nd Law in both the horizontal and vertical directions to solve for the electrical force.

$$F_{net_X} = T\sin\theta - F_e = 0 \rightarrow T\sin\theta = F_e$$

$$F_{net_Y} = T\cos\theta - mg = 0 \rightarrow T\cos\theta = mg$$

$$\frac{T\sin\theta}{T\cos\theta} = \frac{F_e}{mg} \rightarrow \tan\theta = \frac{F_e}{mg} \rightarrow F_e = mg\tan\theta$$

Finally, utilize Coulomb's Law to solve for the charge on each of the balls.

$$F_e = mg\tan\theta \rightarrow \frac{kq_1q_2}{r^2} = mg\tan\theta \xrightarrow[q_1=q_2=q]{r=2L\sin\theta} \frac{kq^2}{4L^2\sin^2\theta} = mg\tan\theta \rightarrow$$

$$q = \sqrt{\frac{4mgL^2\sin^2\theta\tan\theta}{k}} = \sqrt{\frac{4(.005)(9.8)(0.2)^2\sin^2(6°)\tan(6°)}{(8.99\times10^9)}} \rightarrow$$

$$q = 3.16\times10^{-8}C$$

5.28 Q: A +4 C charge and a -2C charge are situated on a number line as shown in the diagram below. Where on the number line is there no net force on a positive charge?

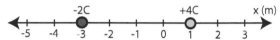

5.28 A: First recognize that the only place on the number line where the net force can be zero is to the left of the -2C charge. Call r the distance from the -2C charge to this point. The distance from r to the +4C charge must therefore be r+4. Now apply Coulomb's Law to the situation to solve for r.

$$\frac{kq_1q_2}{r^2} = \frac{kq_1q_3}{(r+4)^2} \rightarrow \frac{(2)(1)}{r^2} - \frac{(4)(1)}{(r+4)^2} \rightarrow \frac{2}{r^2} = \frac{4}{r^2+8r+16} \rightarrow$$

$$2r^2 = r^2+8r+16 \rightarrow r^2-8r-16 = 0 \xrightarrow{quad\_formula} r = 9.66 \text{ or } r = -1.66$$

The negative value of r won't assist in solving the given situation as that would put the position of the positive charge to the right of the -2C charge, resulting in a net force to the left. Therefore, r=9.66 must be the only solution to our problem. In this case, if r=9.66, and r is the distance from the -2C charge to the point at which there is no net force, the x position of the point at which there is no net force must be -3m - 9.66m or x=-12.66m.

Electric Fields

Similar to gravity, the electrostatic force is a non-contact force, or field force. Charged objects do not have to be in contact with each other to exert a force on each other. Somehow, a charged object feels the effect of another charged object through space. The property of space that allows a charged object to feel a force is a concept called the electric field. Although you cannot see an electric field, you can detect its presence by placing a positive test charge at various points in space and measuring the force the test charge feels.

While looking at gravity, the gravitational field strength was the amount of force observed by a mass per unit mass. In similar fashion, the electric field strength is the amount of electrostatic force observed by a charge per unit charge. Therefore, the electric field strength, E, is the electrostatic force observed at a given point in space divided by the test charge itself. Electric field strength is measured in newtons per coulomb (N/C), which are equivalent to volts per meter (V/m).

$$\left|\vec{E}\right| = \left|\frac{\vec{F}_e}{q}\right| = \left|\frac{kq_1 q_2}{qr^2}\right| = \left|\frac{kq}{r^2}\right|$$

5.29 Q: Two oppositely charged parallel metal plates, 1.00 centimeter apart, exert a force with a magnitude of 3.60 × 10⁻¹⁵ newtons on an electron placed between the plates. Calculate the magnitude of the electric field strength between the plates.

5.29 A: $E = \frac{F_e}{q} = \frac{3.6 \times 10^{-15} N}{1.6 \times 10^{-19} C} = 2.25 \times 10^4 \, {}^N\!\!/_C$

5.30 Q: Which quantity and unit are correctly paired?
(A) resistivity and Ω/m
(B) potential difference and eV
(C) current and C•s
(D) electric field strength and N/C

5.30 A: (D) electric field strength and N/C.

5.31 Q: What is the magnitude of the electric field intensity at a point where a proton experiences an electrostatic force of magnitude 2.30×10⁻²⁵ newtons?
(A) 3.68×10⁻⁴⁴ N/C
(B) 1.44×10⁻⁶ N/C
(C) 3.68×10⁶ N/C
(D) 1.44×10⁴⁴ N/C

5.31 A: (B) $E = \frac{F_e}{q} = \frac{2.3 \times 10^{-25} N}{1.6 \times 10^{-19} C} = 1.44 \times 10^{-6} \, {}^N\!\!/_C$

5.32 Q: The diagram below represents an electron within an electric field between two parallel plates that are charged with a potential difference of 40.0 volts.

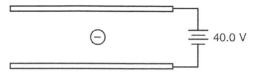

If the magnitude of the electric force on the electron is 2.00×10^{-15} newtons, the magnitude of the electric field strength between the charged plates is

(A) 3.20×10^{-34} N/C

(B) 2.00×10^{-14} N/C

(C) 1.25×10^{4} N/C

(D) 2.00×10^{16} N/C

5.32 A: (C) $E = \dfrac{F_e}{q} = \dfrac{2 \times 10^{-15} N}{1.6 \times 10^{-19} C} = 1.25 \times 10^{4} \, N/_{C}$

Since you can't actually see the electric field, you can draw electric field lines to help visualize the force a charge would feel if placed at a specific position in space. These lines show the direction of the electric force a positively charged particle would feel at that point. The more dense the lines are, the stronger the force a charged particle would feel, therefore the stronger the electric field. As the lines get further apart, the strength of the electric force a charged particle would feel is smaller, therefore the electric field is smaller.

By convention, electric field lines are drawn showing the direction of force on a positive charge. Therefore, to draw electric field lInes for a system of charges, follow these basic rules:

1. Electric field lines point away from positive charges and toward negative charges.
2. Electric field lines never cross.
3. Electric field lines always intersect conductors at right angles to the surface.
4. Stronger fields have closer lines.
5. Field strength and line density decreases as you move away from the charges.

Let's take a look at a few examples of electric field lines, starting with isolated positive (left) and negative (right) charges. Notice that for each charge, the lines radiate outward or inward spherically. The lines point away from the positive charge, since a positive test charge placed in the field (near the fixed charge) would feel a repelling force. The lines point in toward the negative fixed charge, since a positive test charge would feel an attractive force.

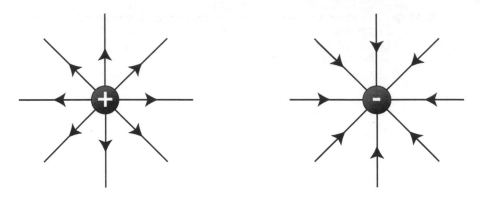

If you have both positive and negative charges in close proximity, you follow the same basic procedure:

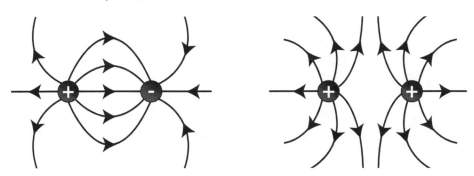

5.33 Q: Two small metallic spheres, A and B, are separated by a distance of 4.0×10^{-1} meter, as shown. The charge on each sphere is $+1.0 \times 10^{-6}$ coulomb. Point P is located near the spheres.

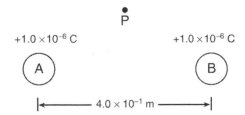

What is the magnitude of the electrostatic force between the two charged spheres?

(A) 2.2×10^{-2} N

(B) 5.6×10^{-2} N

(C) 2.2×10^{4} N

(D) 5.6×10^{4} N

5.33 A: (B) $F_e = \dfrac{kq_1 q_2}{r^2} = \dfrac{(8.99 \times 10^9 \frac{N \bullet m^2}{C^2})(1.0 \times 10^{-6} C)(1.0 \times 10^{-6} C)}{(4 \times 10^{-1} m)^2}$

$F_e = 0.056 N$

5.34 Q: In the diagram below, P is a point near a negatively charged sphere.

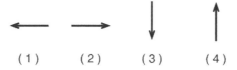

Which vector best represents the direction of the electric field at point P?

(1)   (2)   (3)   (4)

5.34 A: (1) Electric field lines point in toward negative charges.

5.35 Q: Sketch at least four electric field lines with arrowheads that represent the electric field around a negatively charged conducting sphere.

5.35 A:

5.36 Q: The centers of two small charged particles are separated by a distance of 1.2×10^{-4} meter. The charges on the particles are $+8.0 \times 10^{-19}$ coulomb and $+4.8 \times 10^{-19}$ coulomb, respectively. Sketch at least four electric field lines in the region between the two positively charged particles.

5.36 A:

Note that these field lines are not symmetrical since the charges have differing magnitudes.

5.37 Q: Which graph best represents the relationship between the magnitude of the electric field strength, E, around a point charge and the distance, r, from the point charge?

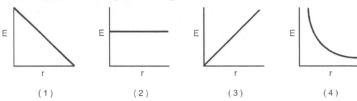

(1)   (2)   (3)   (4)

5.37 A: (4)

Because gravity and electrostatics have so many similarities, let's take a minute to do a quick comparison of electrostatics and gravity.

Electrostatics	Gravity
Force: $F_e = \dfrac{kq_1q_2}{r^2}$	**Force:** $F_g = \dfrac{Gm_1m_2}{r^2}$
Field Strength: $E = \dfrac{F_e}{q}$	**Field Strength:** $g = \dfrac{F_g}{m}$
Field Strength: $E = \dfrac{kq}{r^2}$	**Field Strength:** $g = \dfrac{Gm}{r^2}$
Constant: $k=8.99\times10^9$ N·m²/C²	**Constant:** G=6.67×10⁻¹¹ N·m²/kg²
Charge Units: coulombs	**Mass Units:** kilograms

What is the big difference between electrostatics and gravity? The gravitational force can only attract, while the electrostatic force can both attract and repel. Notice again that both the electric field strength and the gravitational field strength follow the inverse-square law relationship. Field strength is inversely related to the square of the distance.

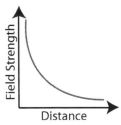

5.38 Q: Determine the x-coordinate where the electric field is zero using the diagram below.

5.38 A: The total electric field is the sum of the electric fields due to the two separate charges. By inspection, you can see that the point where the electric field is zero is between the two charges. Call r the distance from the +1C charge to the point where the electric field is zero. The distance from the +2C charge to the point where the field zero is therefore 11-r.

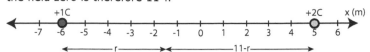

Use the equation for the electric field strength due to a point charge to solve for the distance r.

$$E_{tot} = E_1 + E_2 = \frac{kq_1}{r^2} - \frac{kq_2}{(11-r)^2} = 0 \rightarrow$$

$$\frac{1}{r^2} = \frac{2}{121-22r+r^2} \rightarrow 2r^2 = 121-22r+r^2 \rightarrow$$

$$r^2 + 22r - 121 = 0 \xrightarrow{\quad quad\_formula \quad} r = 4.56 \text{ or } r = -26.6$$

Based on our previous estimation, we can eliminate r=-26.6 as an answer as it would lead to a position that is not between the two positive charges. Therefore, the x-position where there is no electric field can be found from x=-6m + 4.56m = -1.44m.

5.39 Q: Find the electric field at the origin due to the three charges shown in the diagram below.

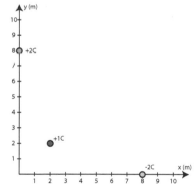

5.39 A: To find the total electric field at the origin, first find the electric field at the origin due to each of the three point charges.

$$E_{+2C} = \frac{kq}{r^2} = \frac{(9 \times 10^9)(2)}{8^2} = -2.81 \times 10^8 \, {}^N\!/_C \, \hat{j} = <0, 2.81 \times 10^8 > {}^N\!/_C$$

$$E_{-2C} = \frac{kq}{r^2} = \frac{(9 \times 10^9)(2)}{8^2} = 2.81 \times 10^8 \, {}^N\!/_C \, \hat{i} = <2.81 \times 10^8, 0 > {}^N\!/_C$$

$$E_{+1C} = \frac{kq}{r^2} = \frac{(9 \times 10^9)(1)}{(2\sqrt{2})^2} = 1.13 \times 10^9 \, {}^{N}\!/_{C} \swarrow \rightarrow$$

$$E_{+1C} = <1.13 \times 10^9 \cos 45°, -1.13 \times 10^9 \sin 45°> {}^{N}\!/_{C} \rightarrow$$

$$E_{+1C} = <-7.95 \times 10^8, -7.95 \times 10^8> {}^{N}\!/_{C}$$

The vector sum of the electric fields due to each of the individual charges gives the total electric field.

$$E_{tot} = <2.81 \times 10^8 - 7.95 \times 10^8, -2.81 \times 10^8 - 7.95 \times 10^8> {}^{N}\!/_{C} \rightarrow$$

$$E_{tot} = <-5.14 \times 10^8, -1.08 \times 10^9> {}^{N}\!/_{C}$$

5.40 Q: The diagrams below each show a point and a charge fixed in space. Rank the magnitude of the electric field at the point from greatest to least for each situation.

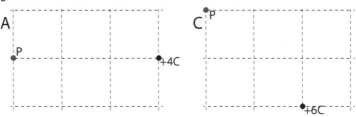

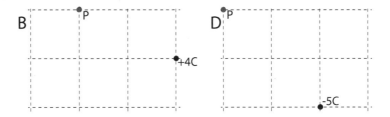

5.40 A: B>C>D>A

The magnitude of the electric field strength is proportional to the charge divided by the square of the distance from the charge.

5.41 Q: A positively charged particle is fixed at point P in space, a set distance d from a neutral solid sphere, as shown below. In Scenario A, the sphere is an insulator. In Scenario B, the sphere is a conductor. Which of the following best describes the forces on the sphere due to the charged particle?

Scenario A

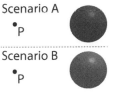

•P

Scenario B

•P

(A) In both scenarios the spheres are attracted to the particle.

(B) In both scenarios the spheres are repelled from the particle.

(C) In scenario A the sphere is attracted and in scenario B the sphere is repelled.

(D) In scenario A the sphere is repelled and in scenario B the sphere is attracted.

5.41 A: (A) In both scenarios the spheres are attracted to the particle. The positively charged particle induces a net negative charge on the closest side of the sphere in both scenarios, resulting in a net force of attraction. There is also an induced positive charge on the far side of the spheres, though the resulting force of repulsion is less than the force of attraction from the near side charges due to the greater separation.

5.42 Q: A sphere with charge -q is placed at rest in an electric field as shown. For each change listed below, describe the effect on the system by filling in the table.

$\vec{E}$

Change	Direction of E field	Magnitude of force on sphere	Direction of sphere's accel.	Magnitude of acceleration
Sign of charge is switched				
E field direction is opposite				
Charge is doubled				
E field is quartered				

5.42 A:

Change	Direction of E field	Magnitude of force on sphere	Direction of sphere's accel.	Magnitude of acceler-ation
Sign of charge is switched	Same	Same	Opposite	Same
E field direction is opposite	Opposite	Same	Opposite	Same
Charge is doubled	Same	Doubled	Same	Doubled
E field is quartered	Same	Quartered	Same	Quartered

5.43 Q: Charged particles are arranged as shown at the corners of equilateral triangles which all have the same area. Rank the magnitude of the electric field strength at point P in each triangle from greatest to least.

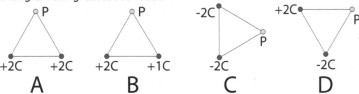

5.43 A: A=C, B, D

The magnitude of the electric field at point P will be maximum for triangles A and C as the two-coulomb charges at a distance of one side length from the point of interest create electric fields which superimpose to create the largest net field as they point largely in the same direction. Triangle B will be next as the superposition of the fields due to the individual charges still creates a net gain in electric field magnitude, while triangle D has the least electric field strength as the electric field from each of the charges point mostly in opposite directions, leaving a minimum resultant electric field.

Although this can be solved quantitatively, a quick graphical analysis on scrap paper with scaled vectors is a quick and efficient method of solving this problem.

5.44 Q: A positive one-coulomb charge is situated at one corner of a cube of volume 1 cubic meter. Determine the magnitude of the electric field strength at the opposite corner of the cube.

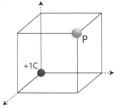

5.44 A: First find the distance between the charge and the point of interest P. This can be done by first finding the diagonal across the base of the cube using the Pythagorean Theorem. If the volume of the cube is 1 m³, each side of the cube must be 1 m. The diagonal across the base of the cube is therefore √2 m. Then, find the diagonal of the cube again using the Pythagorean Theorem:

$$r = \sqrt{\left(\sqrt{2}m\right)^2 + \left(1m\right)^2} = \sqrt{3}m$$

Next, solve for the electric field strength at that distance:

$$\left|\vec{E}\right| = \frac{1}{4\pi\varepsilon_0} \frac{|q|}{r^2} = \frac{1}{4\pi\varepsilon_0} \frac{|1C|}{\left(\sqrt{3}m\right)^2} = 3\times10^9 \ {}^{N}\!/_{C}$$

5.45 Q: A point charge is situated at the center of a series of concentric circles of radius r, 2r, 3r, etc. If the magnitude of the electric field strength at point B is E, determine the magnitude of the electric field strength at points A, C, and D.

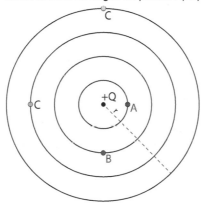

5.45 A: $E_A=4E$; $E_C=4E/9$; $E_D=E/4$

The electric field at each point is inversely proportional to the square of the distance from the charge.

$$E_B = \frac{kQ}{(2r)^2} = \frac{1}{4}\frac{kQ}{r^2} = E$$

$$E_A = \frac{kQ}{r^2} = 4E$$

$$E_C = \frac{kQ}{\left(3r\right)^2} = \frac{kQ}{9r^2} = \frac{4}{9}E$$

$$E_D = \frac{kQ}{\left(4r\right)^2} = \frac{kQ}{16r^2} = \frac{1}{4}E$$

Electric Potential Difference

When an object was lifted against the force of gravity by applying a force over a distance, work was done to give that object gravitational potential energy. The same concept applies to electric fields. If you move a charge against an electric field, you must apply a force for some distance. Therefore, you do work and give it electrical potential energy. The work done per unit charge in moving a charge between two points in an electric field is known as the **electric potential difference**, (V), or electric potential. The units of electric potential are volts, where a volt is equal to 1 joule per coulomb. Therefore, if you do 1 joule of work in moving a charge of 1 coulomb in an electric field, the electric potential difference between those points would be 1 volt. This is described mathematically by:

$$\Delta U_E = q\Delta V$$

V in this formula is potential difference (in volts), ΔU_E is work or change in electrical energy (in joules), and q is your charge (in coulombs). Let's take a look at some sample problems.

5.46 Q: A potential difference of 10 volts exists between two points, A and B, within an electric field. What is the magnitude of charge that requires 2.0×10^{-2} joules of work to move it from A to B?

5.46 A: $\Delta U_E = q\Delta V \rightarrow q = \dfrac{\Delta U_E}{\Delta V} = \dfrac{2\times10^{-2}J}{10V} = 2\times10^{-3}C$

5.47 Q: How much electrical energy is required to move a 4.00-microcoulomb charge through a potential difference of 36.0 volts?

(A) 9.00×10^6 J

(B) 144 J

(C) 1.44×10^{-4} J

(D) 1.11×10^{-7} J

5.47 A: (C) $\Delta U_E = q\Delta V = (4\times10^{-6}C)(36V) = 1.44\times10^{-4}J$

5.48 Q: If 1.0 joule of work is required to move 1.0 coulomb of charge between two points in an electric field, the potential difference between the two points is

(A) 1.0×10^0 V

(B) 9.0×10^9 V

(C) 6.3×10^{18} V

(D) 1.6×10^{-19} V

5.48 A: (A) $\Delta U_E = q\Delta V \rightarrow \Delta V = \dfrac{\Delta U_E}{q} = \dfrac{1J}{1C} = 1V = 1.0 \times 10^0 V$

5.49 Q: Five coulombs of charge are moved between two points in an electric field where the potential difference between the points is 12 volts. How much work is required?

(A) 5 J

(B) 12 J

(C) 60 J

(D) 300 J

5.49 A: (C) $\Delta U_E = q\Delta V = (5C)(12V) = 60J$

5.50 Q: In an electric field, 0.90 joules of work is required to bring 0.45 coulombs of charge from point A to point B. What is the electric potential difference between points A and B?

(A) 5.0 V

(B) 2.0 V

(C) 0.50 V

(D) 0.41 V

5.50 A: (B) $\Delta U_E = q\Delta V \rightarrow \Delta V = \dfrac{\Delta U_E}{q} = \dfrac{0.9J}{0.45C} = 2V$

When dealing with electrostatics, often times the amount of electric energy or work done on a charge is a very small portion of a joule. Dealing with such small numbers is cumbersome, so physicists devised an alternate unit for electrical energy and work that can be more convenient than the joule. This unit, known as the electronvolt (eV), is the amount of work done in moving an elementary charge through a potential difference of 1V. One electron-volt, therefore, is equivalent to one volt multiplied by one elementary charge (in coulombs): 1 eV = 1.6×10^{-19} joules.

5.51 Q: A proton is moved through a potential difference of 10 volts in an electric field. How much work, in electronvolts, was required to move this charge?

5.51 A: $\Delta U_E = q\Delta V = (1e)(10V) = 10eV$

Isolines can be a valuable tool for visualizing electric fields and electric potentials. On a topographic map showing elevation, isolines, also known as contour lines, show regions of equal elevation, which correspond to regions of equal gravitational potential energy for a given mass. These lines can be thought of as lines of gravitational equipotential. Note the lines of equal elevation depicted in the topographic map of Hawaii Island shown below. The distance between the isolines indicates the steepness of the slope. Smaller distances between isolines represent a greater change in elevation per horizontal distance, and are therefore steeper than regions which have more distance between isolines.

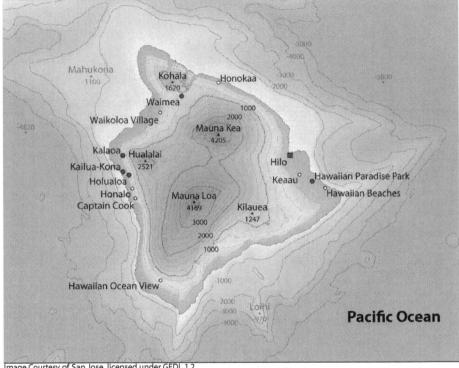

Image Courtesy of San Jose, licensed under GFDL 1.2

In similar fashion, you can analyze regions of an electric field by drawing a map showing both electric field lines and isolines representing lines of equal electric potential, known as equipotential lines. Equipotential lines always intersect electric field lines at right angles. This means that the net electric field along an equipotential line is always zero. Put another way, no work is done on a charge which moves through space while staying on an equipotential line. The figure at right shows the electric field lines pointing radially outward from a positive point charge, with the equipotential lines V_1 and V_2 intersecting the electric field lines at right angles.

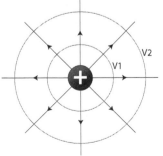

Looking at this concept from another perspective, the electric field vector points from high potential to low potential, and runs perpendicular to any equipotential lines it crosses. In this way, you can state that the electric field is related to the steepness of the change in electric potential (similar to the steepness of a hill on a topographic map). Mathematically, this can be stated as the electric field is equal to the opposite of the gradient of electric potential, where the gradient is the rate at which the electric potential changes with respect to a change in the distance. Limiting ourselves to one dimension, this means the slope of the voltage vs. position curve is the opposite of the electric field strength.

The electric field, then, is the change in electric potential per amount of displacement. Mathematically, this is written as:

$$\vec{E} = -\frac{\Delta V}{\Delta r}$$

For the AP exam, the equation sheet focuses on the magnitude of the electric field strength, so you will see the formula written as:

$$|\vec{E}| = \left|\frac{\Delta V}{\Delta r}\right|$$

5.52 Q: A plot of electric potential difference vs. position along a line is shown below. Use the plot to answer the following questions.

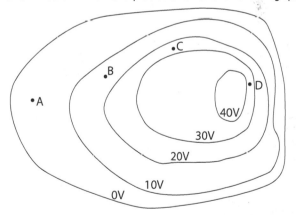

A) Where is the magnitude of the electric field strength greatest?

B) What is the magnitude of the electric field strength at point A?

C) What is the speed of a 10 micro-gram particle released from rest on the 30V equipotential when it reaches the 10V equipotential if its charge is 1 mC?

5.52 A: A) D has the strongest electric field strength since the contour lines are closest together.

B) At A, the contour lines are roughly 1 cm (0.01m) apart, therefore the electric field strength is 10V/0.01m = 1000 V/m.

C) The change in electric potential energy will be equal in magnitude to the change in the particle's kinetic energy, therefore:

$$\Delta U_E = q\Delta V \rightarrow \tfrac{1}{2}mv^2 = q\Delta V \rightarrow v = \sqrt{2\frac{(10^{-3}C)(20V)}{(1\times10^{-8}C)}} = 2000 \,m/_s$$

5.53 Q: A plot of electric potential vs. x-position is shown in the diagram below.

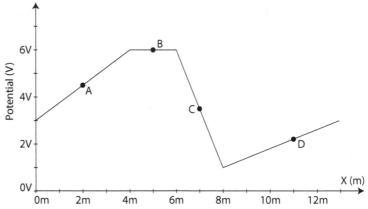

Use the diagram to answer the following questions:

A) At which point(s) is the electric field zero?

B) Where is the magnitude of the electric field strength at a maximum?

C) What is the electric field strength at Point A?

5.53 A: A) Point B. The electric field strength can be found by taking the opposite of the slope of the potential vs. position graph.

B) Point C. The slope of the potential vs. position graph is steepest at Point C.

C) $E = -slope = -\dfrac{rise}{run} = -\dfrac{(3V)}{(4m)} = -0.75 \,V/_m$

5.54 Q: Based on the plot of equipotentials below, which is drawn to scale, answer the following questions:

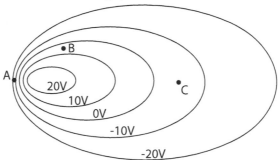

A) Rank the magnitude of the electric field strength at points A, B, and C from greatest to least.

B) Determine the magnitude of the electric field strength at point C.

C) Draw the direction of the electric field at points A, B, and C.

5.54 A: A) A, B, C

B) $|\vec{E}| = \left|\dfrac{\Delta V}{\Delta r}\right| \approx \dfrac{10V}{1.35cm} = \dfrac{10V}{0.0135m} = 740\,V\!/\!_m$

C) The electric field runs from high potential to low potential, and electric field lines intersect equipotential lines at right angles.

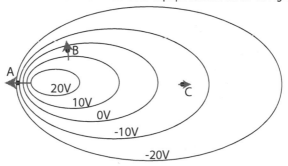

Charges in space create electric fields; therefore charges in space create regions of electric potential. You can calculate the electric potential due to a point charge from the following relationship:

$$V = \frac{1}{4\pi\varepsilon_0}\frac{q}{r} = \frac{kq}{r}$$

For more than one point charge, you can determine the electric potential at a point by adding up the potentials at that point from each individual charge, and because electric potential is a scalar, you no longer need to worry about direction, simplifying such calculations considerably. Electrical potential due to a series of point charges can be determined from:

$$V = \sum \frac{1}{4\pi\varepsilon_0}\frac{q}{r} = \sum \frac{kq}{r} = k\sum \frac{q}{r}$$

5.55 Q: The diagram below shows four point charges of magnitude Q fixed in space, along with four specific points in space, labeled A through D.

A) At which of the four points A through D, if any, is the electric field strength zero?

B) At which of the four points, A through D, if any, is the electric potential zero?

5.55 A: A) None of the points. There is a net electric field at each point A, B, C, and D.

B) Points B & C. The sum of the potentials due to each individual point charge is zero at B and C since each positive charge is offset by an equal charge of opposite magnitude at the same separation distance at those points.

5.56 Q: The diagram below shows two point charges of charge +Q fixed in space, along with five specific points in space, labeled A through E.

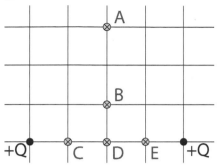

A) Rank the magnitude of the electric field strength at points A through E from greatest to least.

B) Rank the magnitude of the electric potential at points A through E from greatest to least.

5.56 A: A) C=E, B, A, D

B) C=E, D, B, A

5.57 Q: Find the electric potential at the origin due to the following charges: 2µC at (3,0); -5µC at (0,5); and 1µC at (4,4).

5.57 A: $$V = \frac{1}{4\pi\varepsilon_0}\sum\frac{q}{r} = \frac{1}{4\pi\varepsilon_0}\left(\frac{2\times10^{-6}}{3} + \frac{-5\times10^{-6}}{5} + \frac{1\times10^{-6}}{\sqrt{4^2+4^2}}\right) = -1410V$$

5.58 Q: Two point charges (5µC and 2µC) are placed 0.5 meters apart.

A) How much work was required to establish the charge system?

B) What is the electric potential halfway between the two charges?

5.58 A: A) $$U_e = qV = \frac{1}{4\pi\varepsilon_0}\frac{q_1q_2}{r} = \frac{1}{4\pi\varepsilon_0}\frac{(5\times10^{-6})(2\times10^{-6})}{0.5} = 0.18J$$

B) $$V = \frac{1}{4\pi\varepsilon_0}\sum\frac{q}{r} = \frac{1}{4\pi\varepsilon_0}\left(\frac{5\times10^{-6}}{.25} + \frac{2\times10^{-6}}{.25}\right) = 252kV$$

5.59 Q: Two charges are placed along a line as shown in the diagram below.

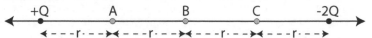

+Q   A   B   C   -2Q

A) Rank the electric field strength from highest to lowest at points A, B, and C.

B) Rank the electric potential from highest to lowest at points A, B, and C.

5.59 A: A) Electric Field: C, A, B

$$E_A = \frac{kQ}{r^2} + \frac{k(2Q)}{3r^2} = 1.22\frac{kQ}{r^2}$$

$$E_B = \frac{kQ}{(2r)^2} + \frac{k(2Q)}{(2r)^2} = 0.75\frac{kQ}{r^2}$$

$$E_C = \frac{k(2Q)}{r^2} + \frac{kQ}{(3r)^2} = 2.11\frac{kQ}{r^2}$$

B) Electric Potential: A, B, C

$$V_A = \frac{kQ}{r} - \frac{2kQ}{3r} = 0.33\frac{kQ}{r}$$

$$V_B = \frac{kQ}{2r} - \frac{2kQ}{2r} = -0.5\frac{kQ}{r}$$

$$V_C = \frac{kQ}{3r} - \frac{2kQ}{r} = -1.67\frac{kQ}{r}$$

5.60 Q: The diagram at right shows three identical point charges of charge +Q fixed in space at the corners of an equilateral triangle of side length l. What is the electric potential energy of a charge +q located at point P, the center of the triangle?

5.60 A: First find the separation between the point P and each of the individual charges. You can do this by breaking up the triangle into a smaller triangular subsection, highlighted in the diagram below.

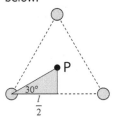

The base of this highlighted right triangle is l/2; therefore, you can use trigonometry to find the hypotenuse of the triangle, which is the separation between the fixed charges and point P.

$$\cos\theta = \frac{adj}{hyp} \rightarrow hyp = \frac{adj}{\cos\theta} = \frac{l/2}{\cos(30°)} = \frac{l}{\sqrt{3}}$$

Now solve for the electric potential at point P.

$$V_P = \sum \frac{kQ}{r} = 3\frac{kQ}{r} = 3\frac{kQ}{l/\sqrt{3}} = 3\sqrt{3}\frac{kQ}{l}$$

The potential energy at point P can then be obtained by multiplying the electric potential at point P by the charge located at that point.

$$U_e = qV = 3\sqrt{3}\frac{kqQ}{l}$$

5.61 Q: The diagram below left depicts the internal structure of a system consisting of three identical point charges +Q with total internal electrical potential energy U. The internal structure of the system changes as depicted in the diagram below right.

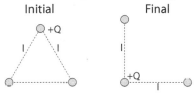

Initial   Final

What is the change in the system's internal energy?

(A) The system's internal energy increases by 10%.

(B) The system's internal energy increases by 15%.

(C) The system's internal energy decreases by 10%.

(D) The system's internal energy decreases by 15%.

5.61 A: (C) The system's internal energy decreases by 10%.

First find the initial electric potential energy of the system. The energy required to bring the first +Q charge from infinity to its current position is 0. The energy required (the work performed) to bring the second +Q charge from infinity to its position at a distance l from the first charge is:

$$W_2 = QV_1 = Q\frac{kQ}{l} = \frac{kQ^2}{l}$$

Then, find the energy required to bring the final +Q charge from infinity to its position a distance l from each of the first two charges.

$$W_3 = QV_1 + QV_2 = Q\frac{kQ}{l} + Q\frac{kQ}{l} = \frac{2kQ^2}{l}$$

The total energy required to assemble this charge configuration is the sum of these energies, which is the electrical potential energy of the configuration.

$$W_{total_{initial}} = \frac{kQ^2}{l} + \frac{2kQ^2}{l} = \frac{3kQ^2}{l}$$

Next, find the electric potential energy of the system in its final configuration in the same way.

$$W_2 = QV_1 = Q\frac{kQ}{l} = \frac{kQ^2}{l}$$

$$W_3 = QV_1 + QV_2 = Q\frac{kQ}{l} + Q\frac{kQ}{\sqrt{l^2 + l^2}} = \frac{kQ^2}{l}\left(1 + \frac{1}{\sqrt{2}}\right)$$

$$W_{total_{initial}} = \frac{kQ^2}{l} + \frac{kQ^2}{l}\left(1 + \frac{1}{\sqrt{2}}\right) = \frac{kQ^2}{l}\left(2 + \frac{1}{\sqrt{2}}\right) \approx 2.707\frac{kQ^2}{l}$$

The change in internal energy, then, is the final value subtracted from the initial value.

$$\Delta U_e = U_{e_{final}} - U_{e_{initial}} = 2.707\frac{kQ^2}{l} - 3\frac{kQ^2}{l} = -0.293\frac{kQ^2}{l}$$

The percent change, then, can be determined to provide the final answer.

$$\frac{\Delta U_e}{U_{e_{initial}}} = \frac{-0.293\frac{kQ^2}{l}}{3\frac{kQ^2}{l}} \approx -10\%$$

Parallel Plates

If you know the potential difference between two parallel plates, you can easily calculate the electric field strength between the plates. As long as you're not near the edge of the plates, the electric field is constant between the plates and its strength is given by the equation:

$$|\vec{E}| = \left|\frac{\Delta V}{\Delta r}\right|$$

You'll note that with the potential difference V in volts, and the distance between the plates in meters, units for the electric field strength are volts per meter [V/m]. Previously, the units for electric field strength were given as newtons per Coulomb [N/C]. It is easy to show these are equivalent:

$$\frac{N}{C} = \frac{N \bullet m}{C \bullet m} = \frac{J}{C \bullet m} = \frac{J/C}{m} = \frac{V}{m}$$

5.62 Q: The magnitude of the electric field strength between two oppositely charged parallel metal plates is 2.0×10^3 newtons per coulomb. Point P is located midway between the plates.

A) Sketch at least five electric field lines to represent the field between the two oppositely charged plates.

B) An electron is located at point P between the plates. Calculate the magnitude of the force exerted on the electron by the electric field.

5.62 A: A)

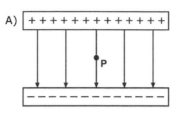

B) $E = \dfrac{F_e}{q}$

$F_e = qE = (1.6 \times 10^{-19} C)(2 \times 10^3 \, {}^N\!/_C) = 3.2 \times 10^{-16} N$

5.63 Q: A moving electron is deflected by two oppositely charged parallel plates, as shown in the diagram below.

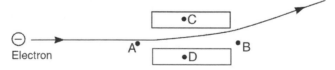

The electric field between the plates is directed from

(A) A to B

(B) B to A

(C) C to D

(D) D to C

5.63 A: (C) C to D because the electron feels a force opposite the direction of the electric field due to its negative charge.

5.64 Q: An electron is located in the electric field between two parallel metal plates as shown in the diagram below.

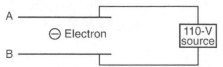

If the electron is attracted to plate A, then plate A is charged
(A) positively, and the electric field runs from plate A to plate B
(B) positively, and the electric field runs from plate B to plate A
(C) negatively, and the electric field runs from plate A to plate B
(D) negatively, and the electric field runs from plate B to plate A

5.64 A: (A) positively, and the electric field runs from plate A to plate B

5.65 Q: An electron placed between oppositely charged parallel plates moves toward plate A, as represented in the diagram below.

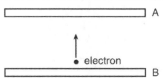

What is the direction of the electric field between the plates?
(A) toward plate A
(B) toward plate B
(C) into the page
(D) out of the page

5.65 A: (B) toward plate B.

5.66 Q: The diagram below represents two electrons, e_1 and e_2, located between two oppositely charged parallel plates.

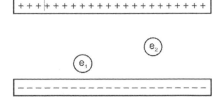

Compare the magnitude of the force exerted by the electric field on e_1 to the magnitude of the force exerted by the electric field on e_2.

5.66 A: The forces are the same because the electric field between two parallel plates is constant.

Chapter 5: Electrostatics

Capacitors

Parallel conducting plates separated by an insulator can be used to store electrical charge. These devices come in a variety of sizes, and are known as parallel plate capacitors. The amount of charge a capacitor can store on a single plate for a given amount of potential difference across the plates is known as the device's capacitance, given in coulombs per volt, also known as a farad (F). A farad is a very large amount of capacitance; therefore, most capacitors have values in the micro-farad, nano-farad, and even pico-farad ranges.

$$C = \frac{Q}{\Delta V} \rightarrow \Delta V = \frac{Q}{C}$$

5.67 Q: A capacitor stores 3 microcoulombs of charge with a potential difference of 1.5 volts across the plates. What is the capacitance?

5.67 A: $C = \dfrac{Q}{\Delta V} = \dfrac{3 \times 10^{-6} F}{1.5V} = 2 \times 10^{-6} F$

5.68 Q: How much charge sits on the top plate of a 200 nF capacitor when charged to a potential difference of 6 volts?

5.68 A: $\Delta V = \dfrac{Q}{C} \rightarrow Q = C \Delta V = (200 \times 10^{-9} F)(6V) = 1.2 \times 10^{-6} C$

The amount of charge a parallel plate capacitor can hold is determined by its geometry as well as the insulating material between the plates. The plates carry equal amounts of charge, but of opposite sign. The capacitance is directly related to the area of the plates, and inversely related to the separation between the plates, as shown in the formula below.

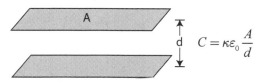

$$C = \kappa \varepsilon_0 \frac{A}{d}$$

The permittivity ($\kappa \varepsilon_0$) of an insulator describes the insulator's resistance to the creation of an electric field. The permittivity of free space, also known as vacuum permittivity, is a constant written as ε_0 and is equal to 8.85×10^{-12} Farads per meter. The symbol kappa (κ) represents the relative permittivity of a material, also known as the dielectric constant. For vacuum and air, the relative permittivity (κ) is equal to 1. For other materials, the relative permittivity is larger. Silicon, for example, has a permittivity of 11.7, while silicon dioxide, a common dielectric material in microprocessors, has a relative permittivity of 3.9.

Dielectrics are insulators which are placed between the plates of a capacitor to increase the device's capacitance. When a dielectric is placed between the plates of a capacitor, the electric field between the plates is weakened due to the molecules of the dielectric becoming polarized in the electric field created by the potential difference of the capacitor plates, creating an opposing electric field. The greater the amount of polarization, the greater the reduction in the electric field. Therefore, for a fixed charge on the plates Q, the voltage decreases, increasing the capacitance (C=Q/V).

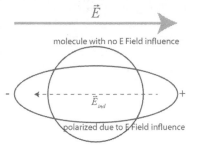

Approached from an alternate perspective, the amount by which the capacitance is increased when a dielectric is introduced between the plates of a capacitor is the relative permittivity, or dielectric constant. This constant also corresponds to the amount the electric field strength is reduced due to the introduction of the dielectric. The more the molecules / atoms of the dielectric are polarized, the greater the dielectric constant.

5.69 Q: Find the capacitance of two parallel plates of length 1 millimeter and width 2 millimeters if they are separated by 3 micrometers of air.

5.69 A: $C = \kappa \varepsilon_0 \dfrac{A}{d} = (1)(8.85 \times 10^{-12}\ ^F\!/_m)\dfrac{(0.001m \times 0.002m)}{3 \times 10^{-6}\, m} = 5.9 \times 10^{-12}\ F$

5.70 Q: Find the distance between the plates of a 5 nano-farad capacitor with a plate area of 0.06 m² and a dielectric of silicon dioxide (relative permittivity = 3.9).

5.70 A: $C = \kappa \varepsilon_0 \dfrac{A}{d} \rightarrow d = \dfrac{\kappa \varepsilon_0 A}{C} = \dfrac{(3.9)(8.85 \times 10^{-12}\ ^F\!/_m)(0.06m^2)}{5 \times 10^{-9}\, F} = 4.1 \times 10^{-4}\, m$

By storing charges on the opposing plates of a parallel-plate capacitor, electrical energy is stored between the plates in the form of the electric field. The electric field outside the plates is zero. The magnitude of the electric field strength between the plates is constant and can be determined through some basic manipulations of previously-developed equations.

$$E = \frac{\Delta V}{r} \xrightarrow{\Delta V = \frac{Q}{C}} E = \frac{Q/C}{d} = \frac{Q}{Cd} = \frac{Qd}{\kappa \varepsilon_0 Ad} \rightarrow E = \frac{Q}{\kappa \varepsilon_0 A}$$

5.71 Q: Rank the magnitude of the electric field strength between the plates of the following parallel-plate capacitors from greatest to least.

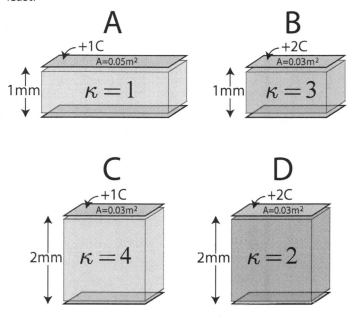

5.71 A: D, B, A, C

5.72 Q: An air-gap parallel-plate capacitor is connected to a source of constant potential difference as shown in the diagram below.

Rank the magnitude of the electric field strength from greatest to least at points A, B, C, D, and E.

5.72 A: B=C=D, A=E. The electric field strength between the plates of the capacitor is constant (assuming you stay away from the edges). The electric field strength outside the capacitor is 0.

The electrical energy stored in the electric field between the plates of the capacitor (U_c) can be quantified as follows:

$$U_c = \frac{1}{2}Q\Delta V = \frac{1}{2}C(\Delta V)^2$$

5.73 Q: Two parallel conducting plates separated by a distance d are connected to a source of constant potential difference. If the distance between the plates is doubled, what happens to the electrical charge on the plates and the electrical potential energy stored in the capacitor?

	charge on plates	potential energy
(A)	halved	halved
(B)	same	halved
(C)	same	same
(D)	doubled	halved
(E)	doubled	same

5.73 A: (A) When the plate separation is doubled, the capacitance is halved; therefore, the charge on the plates is halved (constant potential difference), and the stored electrical potential energy in the capacitor is halved.

5.74 Q: A parallel-plate air-gap capacitor of capacitance C is attached to a constant voltage supply, storing an amount of energy U. The capacitor is then modified in such a way that the area of the plates is halved, the separation between the plates is doubled, and the air-gap is filled with a dielectric of relative permittivity κ=4. What is the amount of energy stored in this new capacitor?

(A) U/4

(B) U/2

(C) U

(D) 2U

(E) 4U

5.74 A: (C) U. The capacitance of the capacitor does not change, and the voltage remains constant, therefore the stored energy does not change.

5.75 Q: An air-gap parallel plate capacitor is charged and then disconnected. While it is disconnected, a dielectric is inserted between the plates. What happens to the stored electric potential energy in the capacitor?

(A) increases

(B) decreases

(C) remains the same

5.75 A: (B) decreases. Charge on the plates remains constant since the capacitor is disconnected. Increasing the dielectric constant betwen the plates increases the capacitance. The electric potential difference between the plates therefore drops ($V=Q/C$). The stored electric potential energy in the capacitor therefore decreases since $U=\frac{1}{2}Q\Delta V$.

5.76 Q: A battery is connected to an air-gap parallel plate capacitor. While the battery is connected, a dielectric is inserted between the plates of the capacitor. Which of the following statements are true? Choose two answers.

(A) The capacitance of the system increases.

(B) The energy stored in the capacitor decreases.

(C) The potential difference across the capacitor increases.

(D) The charge on the plates increases.

(E) The electric field strength between the plates increases.

5.76 A: (A) and (D) are true. The capacitance of the system increases with the increase in permittivity. The energy stored in the capacitor increases, and the potential difference across the capacitor remains constant due to the battery. The charge on the plates increases since the capacitance increases while the voltage remains constant. The electric field strength between the plates remains the same since the potential difference and the plate separation remains constant.

Test Your Understanding

1. Have charges smaller than an elementary charge ever been isolated? Explain.

2. What methods can you use to tell if an object is positively or negatively charged?

3. What conservation laws have you learned? How are they all alike? How are they different?

4. How can you determine if an object is charged?

5. Design an experiment to determine the charge on an object.

6. Explain how contact forces such as tension, friction, normal, buoyant, and spring forces result from interatomic electric forces.

7. Consider the relationship of charge and electric potential difference. Compare and contrast this relationship to the relationship between mass and change in height above the Earth's surface.

8. The electric field inside a conductor at equilibrium is always zero. Explain why.

9. Compare and contrast the use of isolines on a topographic map with the use of isolines in a map of a region of electric field.

10. Why does the net charge in a conductor at equilibrium always reside at the surface of the conductor?

11. Assume each end of a battery is attached to opposite plates of a capacitor. Why does each plate obtain the same magnitude of charge?

Chapter 6: Circuits

"And God said, 'Let there be light' and there was light, but the electricity board said He would have to wait until Thursday to be connected."

— Spike Milligan

Objectives

1. Define and calculate electric current.
2. Define and calculate resistance using Ohm's Law.
3. Explain the factors and calculate the resistance of a conductor.
4. Identify the path and direction of current flow in a circuit.
5. Draw and interpret schematic diagrams of circuits.
6. Effectively use and analyze voltmeters and ammeters.
7. Solve series and parallel circuit problems using VIRP tables.
8. Calculate equivalent resistances for resistors and capacitors in both series and parallel configurations.
9. Calculate power and energy used in electric circuits.
10. Analyze current and voltage in circuits including both series and parallel configurations of basic circuit elements such as resistors and capacitors.

Electric Current

Electric current is the flow of charge, much like water currents are the flow of water molecules. Water molecules tend to flow from areas where they have high gravitational potential energy to areas where they have low gravitational potential energy. Electric charges flow from where they have high electric potential energy to where they have low electric potential energy. The greater the difference between the high and low potential, the more current that flows!

In a majority of electric currents, the moving charges are negative electrons. However, due to historical reasons dating back to Ben Franklin, we say that conventional current flows in the direction positive charges would move. Although inconvenient, it's fairly easy to keep straight if you just remember that the actual moving charges, the electrons, flow in a direction opposite that of the electric current. With this in mind, you can state that positive current flows from high potential to low potential, even though the charge carriers (electrons) actually flow from low to high potential.

Electric current (I) is measured in amperes (A), or amps, and can be calculated by finding the total amount of charge (Δq), in coulombs, which passes a specific point in a given time (t). One ampere is one coulomb per second. Electric current can therefore be calculated as:

$$I = \frac{\Delta q}{t}$$

6.01 Q: A charge of 30 Coulombs passes through a 24-ohm resistor in 6.0 seconds. What is the current through the resistor?

6.01 A: $I = \frac{\Delta q}{t} = \frac{30C}{6s} = 5A$

6.02 Q: Charge flowing at the rate of 2.50×10^{16} elementary charges per second is equivalent to a current of

(A) 2.50×10^{13} A

(B) 6.25×10^5 A

(C) 4.00×10^{-3} A

(D) 2.50×10^{-3} A

6.02 A: (C) $I = \frac{\Delta q}{t} = \frac{(2.50 \times 10^{16})(1.6 \times 10^{-19}C)}{1s} = 4 \times 10^{-3} A$

6.03 Q: The current through a lightbulb is 2.0 amperes. How many coulombs of electric charge pass through the lightbulb in one minute?

(A) 60 C

(B) 2.0 C

(C) 120 C

(D) 240 C

6.03 A: (C) $I = \dfrac{\Delta q}{t}$

$$\Delta q = It = (2A)(60s) = 120C$$

6.04 Q: A 1.5-volt, AAA cell supplies 750 milliamperes of current through a flashlight bulb for 5 minutes, while a 1.5-volt, C cell supplies 750 milliamperes of current through the same flashlight bulb for 20 minutes. Compared to the total charge transferred by the AAA cell through the bulb, the total charge transferred by the C cell through the bulb is

(A) half as great

(B) twice as great

(C) the same

(D) four times as great

6.04 A: (D) If Δq=It, and both cells supply 0.750A but the C cell supplies the same current for four times as long, it must supply four times the total charge compared to the AAA cell.

6.05 Q: The current traveling from the cathode to the screen in a television picture tube is 5.0×10^{-5} amperes. How many electrons strike the screen in 5.0 seconds?

(A) 3.1×10^{24}

(B) 6.3×10^{18}

(C) 1.6×10^{15}

(D) 1.0×10^{5}

6.05 A: (C) $I = \dfrac{\Delta q}{t}$

$$\Delta q = It = (5 \times 10^{-5} A)(5s) = 2.5 \times 10^{-4} C$$

$$2.5 \times 10^{-4} C \bullet \frac{1 \text{ electron}}{1.6 \times 10^{-19} C} = 1.6 \times 10^{15} \text{ electrons}$$

Resistance

Electrical charges can move easily in some materials (conductors) and less freely in others (insulators). A material's ability to conduct electric charge is known as its **conductivity**. Good conductors have high conductivities. The conductivity of a material depends on the material's molecular and atomic structure, specifically:

1. Density of free charges available to move
2. Mobility of those free charges

When an object is created out of a material, the material's tendency to conduct electricity, or conductance, depends on the material's conductivity as well as the material's shape. For example, a hollow cylindrical pipe has a higher conductivity of water than a cylindrical pipe filled with cotton. However, the shape of the pipe also plays a role. A very thick but short pipe can conduct lots of water, yet a very narrow, very long pipe can't conduct as much water. Both geometry of the object and the object's composition influence its conductance.

In similar fashion, material's ability to resist the movement of electric charge is known as its **resistivity**, symbolized with the Greek letter rho (ρ). Resistivity is measured in ohm-meters, which is represented by the Greek letter omega multiplied by meters ($\Omega \bullet m$). Both conductivity and resistivity are properties of a material.

Focusing on an object's ability to resist the flow of electrical charge, objects made of high resistivity materials tend to impede electrical current flow and have a high resistance. Further, materials shaped into long, thin objects also increase an object's electrical resistance. Finally, objects typically exhibit higher resistivities at higher temperatures. You must take all of these factors into account together to describe an object's resistance to the flow of electrical charge. Resistance is a functional property of an object that describes the object's ability to impede the flow of charge through it. Units of resistance are ohms (Ω).

For any given temperature, you can calculate an object's electrical resistance, in ohms, using the following formula.

$$R = \frac{\rho L}{A}$$

Resistivities at 20°C	
Material	**Resistivity ($\Omega \bullet m$)**
Aluminum	2.82×10^{-8}
Copper	1.72×10^{-8}
Gold	2.44×10^{-8}
Nichrome	$150. \times 10^{-8}$
Silver	1.59×10^{-8}
Tungsten	5.60×10^{-8}

In this formula, R is the resistance of the object in ohms (Ω), rho (ρ) is the resistivity of the material the object is made out of in ohm•meters (Ω•m), L is the length of the object in meters, and A is the cross-sectional area of the object in meters squared. Note that a table of material resistivities for a constant temperature is given to you on the previous page as well.

6.06 Q: A 3.50-meter length of wire with a cross-sectional area of 3.14×10^{-6} m^2 at 20° Celsius has a resistance of 0.0625 Ω. Determine the resistivity of the wire and the material it is made out of using the table of resistivities from the previous page.

6.06 A: $$R = \frac{\rho L}{A}$$

$$\rho = \frac{RA}{L} = \frac{(.0625\ )(3.14 \times 10^{-6}\, m^2)}{3.5m} = 5.6 \times 10^{-8}\ \ \bullet m$$

Resistivity of the material is consistent with tungsten.

6.07 Q: The electrical resistance of a metallic conductor is inversely proportional to its

(A) temperature

(B) length

(C) cross-sectional area

(D) resistivity

6.07 A: (C) straight from the formula.

6.08 Q: At 20°C, four conducting wires made of different materials have the same length and the same diameter. Which wire has the least resistance?

(A) aluminum

(B) gold

(C) nichrome

(D) tungsten

6.08 A: (B) gold because it has the lowest resistivity.

6.09 Q: A length of copper wire and a 1.00-meter-long silver wire have the same cross-sectional area and resistance at 20°C. Calculate the length of the copper wire.

6.09 A: $R = \dfrac{\rho L}{A}\Big|_{copper} = \dfrac{\rho L}{A}\Big|_{silver}$

$R = \dfrac{\rho_{copper} L_{copper}}{A} = \dfrac{\rho_{silver} L_{silver}}{A}$

$L_{copper} = \dfrac{\rho_{silver} L_{silver}}{\rho_{copper}} = \dfrac{(1.59 \times 10^{-8}\ m)(1m)}{1.72 \times 10^{-8}\ m}$

$L_{copper} = 0.924m$

6.10 Q: A 10-meter length of copper wire is at 20°C. The radius of the wire is 1.0×10⁻³ meter.

Cross Section of Copper Wire

r = 1.0 × 10⁻³ m

A) Determine the cross-sectional area of the wire.

B) Calculate the resistance of the wire.

6.10 A: A) $Area_{circle} = \pi r^2 = \pi(1.0 \times 10^{-3} m)^2 = 3.14 \times 10^{-6} m^2$

B) $R = \dfrac{\rho L}{A} = \dfrac{(1.72 \times 10^{-8}\ m)(10m)}{3.14 \times 10^{-6} m^2} = 5.5 \times 10^{-2}$

6.11 Q: Which of the following resistors, made of the same material, has the highest resistance?

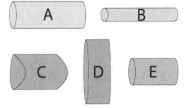

6.11 A: (B) has the highest resistance as it has the greatest length and smallest cross-sectional area.

Ohm's Law

If resistance opposes current flow, and potential difference promotes current flow, it only makes sense that these quantities must somehow be related. Georg Ohm studied and quantified these relationships for conductors and resistors in a famous formula now known as **Ohm's Law**:

$$R = \frac{V}{I}$$

Ohm's Law may make more qualitative sense if it is rearranged slightly:

$$I = \frac{V}{R}$$

Now it's easy to see that the current flowing through a conductor or resistor (in amps) is equal to the potential difference across the object (in volts) divided by the resistance of the object (in ohms). If you want a large current to flow, you require a large potential difference (such as a large battery), and/or a very small resistance.

Note: Ohm's Law isn't truly a law of physics -- not all materials obey this relationship. It is, however, a very useful empirical relationship that accurately describes key electrical characteristics of conductors and resistors. One way to test if a material is ohmic (if it follows Ohm's Law) is to graph the voltage vs. current flow through the material. If the material obeys Ohm's Law, you get a linear relationship, where the slope of the line is equal to the material's resistance.

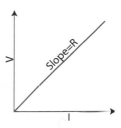

6.12 Q: The current in a wire is 24 amperes when connected to a 1.5 volt battery. Find the resistance of the wire.

6.12 A: $R = \dfrac{V}{I} = \dfrac{1.5V}{24A} = 0.0625$

6.13 Q: In a simple electric circuit, a 24-ohm resistor is connected across a 6-volt battery. What is the current in the circuit?

(A) 1.0 A

(B) 0.25 A

(C) 140 A

(D) 4.0 A

6.13 A: (B) $I = \dfrac{V}{R} = \dfrac{6V}{24} = 0.25A$

6.14 Q: What is the current in a 100-ohm resistor connected to a 0.40-volt source of potential difference?

(A) 250 mA

(B) 40 mA

(C) 2.5 mA

(D) 4.0 mA

6.14 A: (D) $I = \dfrac{V}{R} = \dfrac{0.40V}{100} = 0.004\,A = 4mA$

6.15 Q: A constant potential difference is applied across a variable resistor held at constant temperature. Which graph best represents the relationship between the resistance of the variable resistor and the current through it?

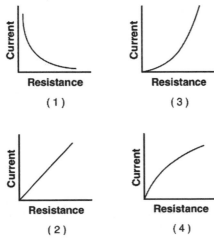

6.15 A: (1) due to Ohm's Law (I=V/R).

6.16 Q: The graph below represents the relationship between the potential difference (V) across a resistor and the current (I) through the resistor.

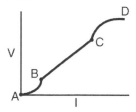

Through which entire interval does the resistor obey Ohm's law?

(A) AB

(B) BC

(C) CD

(D) AD

6.16 A: (B) BC because the graph is linear in this interval.

Electrical Circuits

An **electrical circuit** is a closed loop path through which current can flow. An electrical circuit can be made up of almost any materials (including humans if they're not careful), but practically speaking, circuits are typically comprised of electrical devices such as wires, batteries, resistors, and switches. Conventional current will flow through a complete closed loop (closed circuit) from high potential to low potential. Therefore, electrons actually flow in the opposite direction, from low potential to high potential. If the path isn't a closed loop (and is, instead, an open circuit), no current will flow.

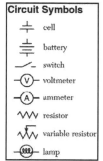

Circuit Symbols	
⊣⊢	cell
⊣⊢	battery
╱	switch
–Ⓥ–	voltmeter
–Ⓐ–	ammeter
⋀⋁⋀	resistor
⋀⋁⋀	variable resistor
–⬭–	lamp

Electric circuits, which are three-dimensional constructs, are typically represented in two dimensions using diagrams known as **circuit schematics**. These schematics are simplified, standardized representations in which common circuit elements are represented with specific symbols, and wires connecting the elements in the circuit are represented by lines. Basic circuit schematic symbols are shown in the diagram at left.

In order for current to flow through a circuit, you must have a source of potential difference. Typical sources of potential difference are voltaic cells, batteries (which are just two or more cells connected together), and power (voltage) supplies. Voltaic cells are often referred to as batteries in common terminology. In drawing a cell or battery on a circuit schematic, remember that the longer side of the symbol is the positive terminal.

Chapter 6: Circuits

Electric circuits must form a complete conducting path in order for current to flow. In the example circuit shown below left, the circuit is incomplete because the switch is open; therefore no current will flow and the lamp will not light. In the circuit below right, however, the switch is closed, creating a closed loop path. Current will flow and the lamp will light up. The greater the power expended by the lamp, the brighter the lamp glows.

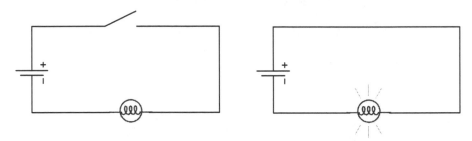

Note that in the picture at right, conventional current will flow from positive to negative, creating a clockwise current path in the circuit. The actual electrons in the wire, however, are flowing in the opposite direction, or counter-clockwise.

Energy & Power

Just like mechanical power is the rate at which mechanical energy is expended, **electrical power** is the rate at which electrical energy is expended. When you do work on something you change its energy. Further, electrical work or energy is equal to charge times potential difference. Therefore, you can combine these to write the equation for electrical power as:

$$P = \frac{W}{t} = \frac{qV}{t}$$

The amount of charge moving past a point per given unit of time is current, and therefore you can continue the derivation as follows:

$$P = \frac{q}{t}\, V = IV$$

So electrical power expended in a circuit is the electrical current multiplied by potential difference (voltage). Using Ohm's Law, you can expand this even further to provide several different methods for calculating electrical power dissipated by a resistor:

$$P = VI = I^2 R = \frac{V^2}{R}$$

Of course, conservation of energy still applies, so the energy used in the resistor is converted into heat (in most cases) and light, or it can be used to do work. Let's put this knowledge to use in a practical application.

6.17 Q: A 110-volt toaster oven draws a current of 6 amps on its highest setting as it converts electrical energy into thermal energy. What is the toaster's maximum power rating?

6.17 A: $P = VI = (110V)(6A) = 660W$

6.18 Q: An electric iron operating at 120 volts draws 10 amperes of current. How much heat energy is delivered by the iron in 30 seconds?
(A) 3.0×10^2 J
(B) 1.2×10^3 J
(C) 3.6×10^3 J
(D) 3.6×10^4 J

6.18 A: (D) $W = Pt = VIt = (120V)(10A)(30s) = 3.6 \times 10^4 J$

6.19 Q: One watt is equivalent to one
(A) N·m
(B) N/m
(C) J·s
(D) J/s

6.19 A: (D) J/s, since Power is W/t, and the unit of work is the joule, and the unit of time is the second.

6.20 Q: A potential drop of 50 volts is measured across a 250-ohm resistor. What is the power developed in the resistor?
(A) 0.20 W
(B) 5.0 W
(C) 10 W
(D) 50 W

6.20 A: (C) $P = \dfrac{V^2}{R} = \dfrac{(50V)^2}{250} = 10W$

6.21 Q: What is the minimum information needed to determine the power dissipated in a resistor of unknown value?

(A) potential difference across the resistor, only

(B) current through the resistor, only

(C) current and potential difference, only

(D) current, potential difference, and time of operation

6.21 A: (C) current and potential difference, only (P=VI).

Voltmeters

Voltmeters are tools used to measure the potential difference between two points in a circuit. The voltmeter is connected in parallel with the element to be measured, meaning an alternate current path around the element to be measured and through the voltmeter is created. You have connected a voltmeter correctly if you can remove the voltmeter from the circuit without breaking the circuit. In the diagram at 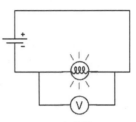 right, a voltmeter is connected to correctly measure the potential difference across the lamp. Voltmeters have very high resistance so as to minimize the current flow through the voltmeter and the voltmeter's impact on the circuit.

Ammeters

Ammeters are tools used to measure the current in a circuit. The ammeter is connected in series with the circuit, so that the current to be measured flows directly through the ammeter. The circuit must be broken to correctly insert an ammeter. Ammeters have very low resistance to minimize the potential drop through the 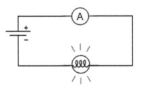 ammeter and the ammeter's impact on the circuit, so inserting an ammeter into a circuit in parallel can result in extremely high currents and may destroy the ammeter. In the diagram at right, an ammeter is connected correctly to measure the current flowing through the circuit.

6.22 Q: In the electric circuit diagram, possible locations of an ammeter and voltmeter are indicated by circles 1, 2, 3, and 4. Where should an ammeter be located to correctly measure the total current and where should a voltmeter be located to correctly measure the total voltage?

Chapter 6: Circuits

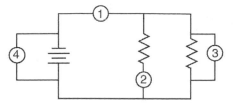

(A) ammeter at 1 and voltmeter at 4

(B) ammeter at 2 and voltmeter at 3

(C) ammeter at 3 and voltmeter at 4

(D) ammeter at 1 and voltmeter at 2

6.22 A: (A) To measure the total current, the ammeter must be placed at position 1, as all the current in the circuit must pass through this wire, and ammeters are always connected in series. To measure the total voltage in the circuit, the voltmeter could be placed at either position 3 or position 4. Voltmeters are always placed in parallel with the circuit element being analyzed, and positions 3 and 4 are equivalent because they are connected with wires (and potential is always the same anywhere in an ideal wire).

6.23 Q: Which circuit diagram below correctly shows the connection of ammeter A and voltmeter V to measure the current through and potential difference across resistor R?

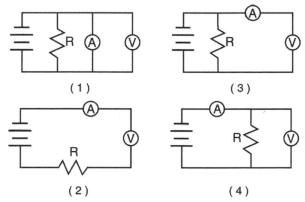

6.23 A: (4) shows an ammeter in series and a voltmeter in parallel with the resistor.

6.24 Q: A student uses a voltmeter to measure the potential difference across a resistor. To obtain a correct reading, the student must connect the voltmeter

(A) in parallel with the resistor

(B) in series with the resistor

(C) before connecting the other circuit components

(D) after connecting the other circuit components

6.24 A: (A) in parallel with the resistor.

6.25 Q: Which statement about ammeters and voltmeters is correct?

(A) The internal resistance of both meters should be low.

(B) Both meters should have a negligible effect on the circuit being measured.

(C) The potential drop across both meters should be made as large as possible.

(D) The scale range on both meters must be the same.

6.25 A: (B) Both meters should have a negligible effect on the circuit being measured.

6.26 Q: Compared to the resistance of the circuit being measured, the internal resistance of a voltmeter is designed to be very high so that the meter will draw

(A) no current from the circuit

(B) little current from the circuit

(C) most of the current from the circuit

(D) all the current from the circuit

6.26 A: (B) the voltmeter should draw as little current as possible from the circuit to minimize its effect on the circuit, but it does require some small amount of current to operate.

Series Circuits

Developing an understanding of circuits is the first step in learning about the modern-day electronic devices that dominate what is becoming known as the "Information Age." A basic circuit type, the **series circuit**, is a circuit in which there is only a single current path. Kirchhoff's Laws are highly effective tools for analyzing these (and other) circuit configurations.

Kirchhoff's Current Law (KCL), named after German physicist Gustav Kirchhoff, states that the sum of all current entering any point in a circuit has to equal the sum of all current leaving any point in a circuit. More simply, this is another way of looking at the law of conservation of charge. This law assumes total electric charge remains constant in the region being considered, a valid assumption in our study of DC circuits.

Kirchhoff's Voltage Law (KVL) states that the sum of all the potential drops in any closed loop of a circuit has to equal zero. More simply, KVL is a method of applying the law of conservation of energy to a circuit. This law assumes there are no changing magnetic fields in the closed loop, also a valid assumption in our study of DC circuits.

6.27 Q: A 3.0-ohm resistor and a 6.0-ohm resistor are connected in series in an operating electric circuit. If the current through the 3.0-ohm resistor is 4.0 amperes, what is the potential difference across the 6.0-ohm resistor?

6.27 A: If 4 amps of current is flowing through the 3-ohm resistor, then 4 amps of current must be flowing through the 6-ohm resistor according to Kirchhoff's Current Law. Since you know the current and the resistance, you can calculate the voltage drop across the 6-ohm resistor using Ohm's Law: $V=IR=(4A)(6\Omega)=24V$.

6.28 Q: The diagram below represents currents in a segment of an electric circuit.

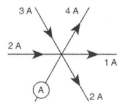

What is the reading of ammeter A?

(A) 1 A

(B) 2 A

(C) 3 A

(D) 4 A

6.28 A: (B) Since five amps plus the unknown current are coming in to the junction, and seven amps are leaving, KCL says that the total current in must equal the total current out; therefore the unknown current must be two amps into the junction.

Let's take a look at a sample circuit, consisting of three 2000-ohm (2 kilo-ohm) resistors:

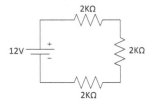

There is only a single current path in the circuit, which travels through all three resistors. Instead of using three separate 2KΩ (2000Ω) resistors, you could replace the three resistors with one single resistor having an equivalent resistance. To find the equivalent resistance of any number of series resistors, just add up their individual resistances:

$$R_{eq} = R_1 + R_2 + R_3 + ...$$
$$R_{eq} = 2000 + 2000 + 2000$$
$$R_{eq} = 6000 = 6K$$

Note that because there is only a single current path, the same current must flow through each of the resistors.

A simple and straightforward method for analyzing circuits involves creating a VIRP table for each circuit you encounter. Combining your knowledge of Ohm's Law, Kirchhoff's Current Law, Kirchhoff's Voltage Law, and equivalent resistance, you can use this table to solve for the details of any circuit.

A VIRP table describes the potential drop (V), current flow (I), resistance (R) and power dissipated (P) for each element in your circuit, as well as for the circuit as a whole. Let's use the circuit with the three 2000-ohm resistors as an example to demonstrate how a VIRP table is used. To create the VIRP table, first list the circuit elements, and total, in the rows of the table, then make columns for V, I, R, and P:

VIRP Table

	V	I	R	P
R₁				
R₂				
R₃				
Total				

Next, fill in the information in the table that is known. For example, you know the total voltage in the circuit (12V) provided by the battery, and you know the values for resistance for each of the individual resistors:

	V	I	R	P
R$_1$			2000Ω	
R$_2$			2000Ω	
R$_3$			2000Ω	
Total	12V			

Once the initial information has been filled in, you can also calculate the total resistance, or equivalent resistance, of the entire circuit. In this case, the equivalent resistance is 6000 ohms.

	V	I	R	P
R$_1$			2000Ω	
R$_2$			2000Ω	
R$_3$			2000Ω	
Total	12V		6000Ω	

Looking at the bottom (total) row of the table, both the voltage drop (V) and the resistance (R) are known. Using these two items, the total current flow in the circuit can be calculated using Ohm's Law.

$$I = \frac{V}{R} = \frac{12V}{6000} = 0.002\,A$$

The total power dissipated can also be calculated using any of the formulas for electrical power.

$$P = \frac{V^2}{R} = \frac{(12V)^2}{6000} = 0.024W$$

More information can now be completed in the VIRP table:

	V	I	R	P
R$_1$			2000Ω	
R$_2$			2000Ω	
R$_3$			2000Ω	
Total	12V	0.002A	6000Ω	0.024W

Because this is a series circuit, the total current has to be the same as the current through each individual element, so you can fill in the current through each of the individual resistors:

	V	I	R	P
R₁		0.002A	2000Ω	
R₂		0.002A	2000Ω	
R₃		0.002A	2000Ω	
Total	12V	0.002A	6000Ω	0.024W

Finally, for each element in the circuit, you now know the current flow and the resistance. Using this information, Ohm's Law can be applied to obtain the voltage drop (V=IR) across each resistor. Power can also be found for each element using P=I²R to complete the table.

	V	I	R	P
R₁	4V	0.002A	2000Ω	0.008W
R₂	4V	0.002A	2000Ω	0.008W
R₃	4V	0.002A	2000Ω	0.008W
Total	12V	0.002A	6000Ω	0.024W

So what does this table really tell you now that it's completely filled out? You know the potential drop across each resistor (4V), the current through each resistor (2 mA), and the power dissipated by each resistor (8 mW). In addition, you know the total potential drop for the entire circuit is 12V, and the entire circuit dissipated 24 mW of power. Note that for a series circuit, the sum of the individual voltage drops across each element equal the total potential difference in the circuit, the current is the same throughout the circuit, and the resistances and power dissipated values add up to the total resistance and total power dissipated. These are summarized for you in the formulas below:

$$I = I_1 = I_2 = I_3 = ...$$
$$V = V_1 + V_2 + V_3 + ...$$
$$R_{eq} = R_1 + R_2 + R_3 + ...$$

6.29 Q: A 2.0-ohm resistor and a 4.0-ohm resistor are connected in series with a 12-volt battery. If the current through the 2.0-ohm resistor is 2.0 amperes, the current through the 4.0-ohm resistor is

(A) 1.0 A

(B) 2.0 A

(C) 3.0 A

(D) 4.0 A

6.29 A: (B) The current through a series circuit is the same everywhere. Therefore, the correct answer must be 2.0 amperes.

6.30 Q: In the circuit represented by the diagram, what is the reading of voltmeter V?

(A) 20 V

(B) 2.0 V

(C) 30 V

(D) 40 V

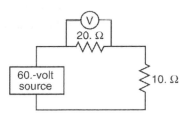

6.30 A: (D) Voltmeter reads potential difference across R_1 which is 40 V.

	V	I	R	P
R_1	40V	2A	20Ω	80W
R_2	20V	2A	10Ω	40W
Total	60V	2A	30Ω	120W

6.31 Q: In the circuit diagram below, two 4.0-ohm resistors are connected to a 16-volt battery as shown.

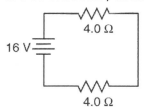

The rate at which electrical energy is expended in this circuit is

(A) 8.0 W

(B) 16 W

(C) 32 W

(D) 64 W

6.31 A: (C) 32W. Rate at which energy is expended is known as power.

	V	I	R	P
R_1	8V	2A	4Ω	16W
R_2	8V	2A	4Ω	16W
Total	16V	2A	8Ω	32W

6.32 Q: A 50-ohm resistor, an unknown resistor R, a 120-volt source, and an ammeter are connected in a complete circuit. The ammeter reads 0.50 ampere.

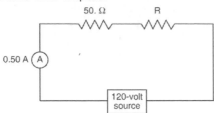

(A) Calculate the equivalent resistance of the circuit.

(B) Determine the resistance of resistor R.

(C) Calculate the power dissipated by resistor R.

6.32 A: (A) R_{eq}= 240Ω   (B) R= 190Ω   (C) $P_{50Ω \ resistor}$= 47.5W

	V	I	R	P
R₁	25V	0.50A	50Ω	12.5W
R₂	95V	0.50A	190Ω	47.5W
Total	120V	0.50A	240Ω	60W

6.33 Q: What must be inserted between points A and B to establish a steady electric current in the incomplete circuit represented in the diagram?

(A) switch

(B) voltmeter

(C) magnetic field source

(D) source of potential difference

6.33 A: (D) a source of potential difference is required to drive current.

Parallel Circuits

Another basic circuit type is the **parallel circuit**, in which there is more than one current path. To analyze resistors in a series circuit, you found an equivalent resistance. You'll follow the same strategy in analyzing resistors in parallel.

Let's examine a circuit made of the same components used in the exploration of series circuits, but now connect the components so as to provide multiple current paths, creating a parallel circuit.

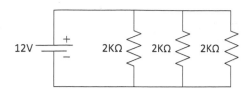

Notice that in this circuit, electricity can follow one of three different paths through the resistors. In many ways, this is similar to a river branching into three different smaller rivers. Each resistor, then, causes a potential drop (analogous to a waterfall), then the three rivers recombine before heading back to the battery, which you can think of like a pump, raising the river to a higher potential before sending it back on its looping path. Or you can think of it as students rushing out of a classroom. The more doors in the room, the less resistance there is to exiting!

You can find the equivalent resistance of resistors in parallel using the formula:

$$\frac{1}{R_{eq}} = \frac{1}{R_1} + \frac{1}{R_2} + \frac{1}{R_3} + \dots$$

Take care in using this equation, as it's easy to make errors in performing your calculations. For only two resistors, this simplifies to:

$$R_{eq} = \frac{R_1 R_2}{R_1 + R_2}$$

Let's find the equivalent resistance for the sample circuit.

$$\frac{1}{R_{eq}} = \frac{1}{R_1} + \frac{1}{R_2} + \frac{1}{R_3} + \dots$$

$$\frac{1}{R_{eq}} = \frac{1}{2000} + \frac{1}{2000} + \frac{1}{2000}$$

$$\frac{1}{R_{eq}} = 0.0015 \, /$$

$$R_{eq} = \frac{1}{0.0015 \, /} = 667$$

A VIRP table can again be used to analyze the circuit, beginning by filling in what is known directly from the circuit diagram.

VIRP Table

	V	I	R	P
R₁			2000Ω	
R₂			2000Ω	
R₃			2000Ω	
Total	12V			

You can also see from the circuit diagram that the potential drop across each resistor must be 12V, since the ends of each resistor are held at a 12-volt difference by the battery

	V	I	R	P
R₁	12V		2000Ω	
R₂	12V		2000Ω	
R₃	12V		2000Ω	
Total	12V			

Next, you can use Ohm's Law to fill in the current through each of the individual resistors since you know the voltage drop across each resistor (I=V/R) to find I=0.006A.

	V	I	R	P
R₁	12V	**0.006A**	2000Ω	
R₂	12V	**0.006A**	2000Ω	
R₃	12V	**0.006A**	2000Ω	
Total	12V			

Using Kirchhoff's Current Law, you can see that if 0.006A flows through each of the resistors, these currents all come together to form a total current of 0.018A.

	V	I	R	P
R₁	12V	0.006A	2000Ω	
R₂	12V	0.006A	2000Ω	
R₃	12V	0.006A	2000Ω	
Total	12V	**0.018A**		

Because each of the three resistors has the same resistance, it only makes sense that the current would be split evenly between them. You can confirm the earlier calculation of equivalent resistance by calculating the total resistance of the circuit using Ohm's Law: R=V/I=(12V/0.018A)=667Ω.

	V	I	R	P
R₁	12V	0.006A	2000Ω	
R₂	12V	0.006A	2000Ω	
R₃	12V	0.006A	2000Ω	
Total	12V	0.018A	667Ω	

Finally, you can complete the VIRP table using any of the three applicable equations for power dissipation to find:

	V	I	R	P
R₁	12V	0.006A	2000Ω	0.072W
R₂	12V	0.006A	2000Ω	0.072W
R₃	12V	0.006A	2000Ω	0.072W
Total	12V	0.018A	667Ω	0.216W

Note that for resistors in parallel, the equivalent resistance is always less than the resistance of any of the individual resistors. The potential difference across each of the resistors in parallel is the same, and the current through each of the resistors adds up to the total current. This is summarized for you in the following table:

$$I = I_1 + I_2 + I_3 + ...$$
$$V = V_1 = V_2 = V_3 = ...$$
$$\frac{1}{R_{eq}} = \frac{1}{R_1} + \frac{1}{R_2} + \frac{1}{R_3} + ...$$

6.34 Q: A 15-ohm resistor, R_1, and a 30-ohm resistor, R_2, are to be connected in parallel between points A and B in a circuit containing a 90-volt battery.

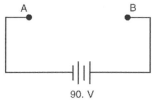

90. V

(A) Complete the diagram to show the two resistors connected in parallel between points A and B.

(B) Determine the potential difference across resistor R_1.

(C) Calculate the current in resistor R_1.

6.34 A: (A)

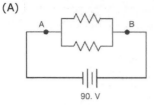

90. V

(B) Potential difference across R_1 is 90V.

(C) Current through resistor R_1 is 6A.

	V	I	R	P
R_1	90V	6A	15Ω	540W
R_2	90V	3A	30Ω	270W
Total	90V	9A	10Ω	810W

6.35 Q: Draw a diagram of an operating circuit that includes: a battery as a source of potential difference, two resistors in parallel with each other, and an ammeter that reads the total current in the circuit.

6.35 A:

6.36 Q: Three identical lamps are connected in parallel with each other. If the resistance of each lamp is X ohms, what is the equivalent resistance of this parallel combination?

(A) X Ω

(B) X/3 Ω

(C) 3X Ω

(D) 3/X Ω

6.36 A: (B) X/3 Ω

$$\frac{1}{R_{eq}} = \frac{1}{R_1} + \frac{1}{R_2} + \frac{1}{R_3} + \dots$$

$$\frac{1}{R_{eq}} = \frac{1}{X} + \frac{1}{X} + \frac{1}{X}$$

$$\frac{1}{R_{eq}} = \frac{3}{X}$$

$$R_{eq} = \frac{X}{3}$$

6.37 Q: Three resistors, 4 ohms, 6 ohms, and 8 ohms, are connected in parallel in an electric circuit. The equivalent resistance of the circuit is

(A) less than 4 Ω

(B) between 4 Ω and 8 Ω

(C) between 10 Ω and 18 Ω

(D) 18 Ω

6.37 A: (A) the equivalent resistance of resistors in parallel is always less than the value of the smallest resistor.

6.38 Q: A 3-ohm resistor, an unknown resistor, R, and two ammeters, A_1 and A_2, are connected as shown with a 12-volt source. Ammeter A_2 reads a current of 5 amperes.

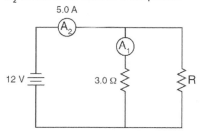

(A) Determine the equivalent resistance of the circuit.

(B) Calculate the current measured by ammeter A_1.

(C) Calculate the resistance of the unknown resistor, R.

6.38 A: (A) 2.4Ω   (B) 4A   (C) 12Ω

	V	I	R	P
R_1	12V	4A	3Ω	48W
R_2	12V	1A	12Ω	12W
Total	12V	5A	2.4Ω	60W

6.39 Q: The diagram below represents an electric circuit consisting of four resistors and a 12-volt battery.

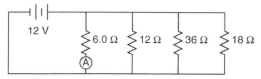

(A) What is the current measured by ammeter A?

(B) What is the equivalent resistance of this circuit?

(C) How much power is dissipated in the 36-ohm resistor?

6.39 A: (A) 2A   (B) 3Ω   (C) 4W

	V	I	R	P
R₁	12V	2A	6Ω	24W
R₂	12V	1A	12Ω	12W
R₃	12V	0.33A	36Ω	4W
R₄	12V	0.67A	18Ω	8W
Total	12V	4A	3Ω	48W

6.40 Q: A 20-ohm resistor and a 30-ohm resistor are connected in parallel to a 12-volt battery as shown. An ammeter is connected as shown.

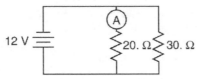

(A) What is the equivalent resistance of the circuit?
(B) What is the current reading of the ammeter?
(C) What is the power of the 30-ohm resistor?

6.40 A: (A) 12Ω   (B) 0.6A   (C) 4.8W

	V	I	R	P
R₁	12V	0.6A	20Ω	7.2W
R₂	12V	0.4A	30Ω	4.8W
Total	12V	1A	12Ω	12W

6.41 Q: In the circuit diagram shown below, ammeter A_1 reads 10 amperes.

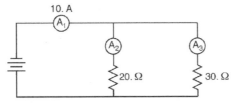

What is the reading of ammeter A_2?
(A) 6 A
(B) 10 A
(C) 20 A
(D) 4 A

6.41 A: (A) 6 A

	V	I	R	P
R₁	120V	6A	20Ω	720W
R₂	120V	4A	30Ω	480W
Total	120V	10A	12Ω	1200W

Batteries

A cell or battery (combination of cells) provides a potential difference, oftentimes referred to as an electromotive force, or emf (ε). A battery can be thought of as a pump for charge, raising charges from a lower potential to a higher potential. Ideal batteries have no resistance, but real batteries have some amount of resistance to the flow of charge within the battery itself, known as the internal resistance (r_i). Because of this internal resistance, the terminal voltage in real batteries is slightly lower than the battery's emf.

In an ideal battery, the terminal voltage, the voltage between points A and B, is the battery's emf. In a real battery, the terminal voltage is the battery's emf minus the voltage drop across the battery's internal resistance.

Ideal Battery

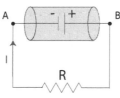

$$V_{battery} = \Delta V = V_B - V_A$$
$$V_{battery} = \varepsilon = V_T$$

Real Battery

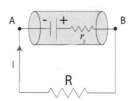

$$V_{battery} = \Delta V = V_B - V_A$$
$$V_{battery} = IR = \varepsilon - Ir_i = V_T$$

6.42 Q: The terminal voltage of a real battery is 15 volts. If the battery has an emf of 18 volts and supplies 10 watts of power to resistor R, find the value of R and r_i.

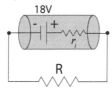

6.42 A: Use a VIRP table to analyze this circuit configuration.

	V	I	R	P
r_i	3 V	0.67 A	4.5 Ω	2 W
R	15 V	0.67 A	22.5 Ω	10 W
Total	18 V	0.67 A	27 Ω	12 W

Combination Series-Parallel Resistor Circuits

A circuit doesn't have to be completely serial or parallel. In fact, most circuits actually have elements of both types. Analyzing these circuits can be accomplished using the fundamentals you learned in analyzing series and parallel circuits separately and applying them in a logical sequence.

First, look for portions of the circuit that have parallel elements. Since the voltage across the parallel elements must be the same, replace the parallel resistors with an equivalent single resistor in series and draw a new schematic. Now you can analyze your equivalent series circuit with a VIRP table. Once your table is complete, work back to your original circuit using KCL and KVL until you know the current, voltage, and resistance of each individual element in your circuit. Note that the columns in VIRP tables for combination circuits are not expected to add up to the total value for any column except power.

6.43 Q: Find the current through R_2 in the circuit below.

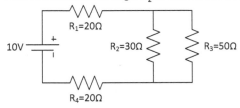

6.43 A: First, find the equivalent resistance for R_2 and R_3 in parallel.

$$R_{eq_{23}} = \frac{R_2 R_3}{R_2 + R_3} = \frac{(30\ )(50\ )}{30\ + 50} = 19$$

Next, re-draw the circuit schematic as an equivalent series circuit.

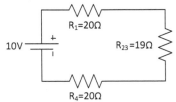

Now, you can use your VIRP table to analyze the circuit.

	V	I	R	P
R_1	3.39V	0.169A	20Ω	0.57W
R_{23}	3.22V	0.169A	19Ω	0.54W
R_4	3.39V	0.169A	20Ω	0.57W
Total	10V	0.169A	59Ω	1.69W

The voltage drop across R_2 and R_3 must therefore be 3.22 volts. From here, you can apply Ohm's Law to find the current through R_2:

$$I_2 = \frac{V_2}{R_2} = \frac{3.22V}{30} = 0.107A$$

	V	I	R	P
R_1	3.39V	0.169A	20Ω	0.57W
R_2	3.22V	0.107A	30Ω	0.34W
R_3	3.22V	0.062A	50Ω	0.20W
R_4	3.39V	0.169A	20Ω	0.57W
Total	10V	0.169A	59Ω	1.69W

6.44 Q: Consider the following four DC circuits.

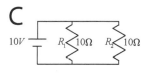
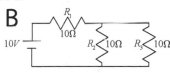
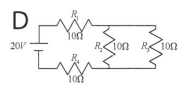

For each circuit, rank the following quantities from highest to lowest in terms of:

I) Current through R_2.

II) Power dissipated by R_1.

III) Equivalent resistance of the entire circuit.

IV) Potential drop across R_2.

I) C, A, D, B

II) C, D, B, A

III) D, A, B, C

IV) C, A, D, B

VIRP Tables for Each Circuit Below:

Circuit A	V	I	R	P
R_1	5V	0.5A	10Ω	2.5W
R_2	5V	0.5A	10Ω	2.5W
Total	10V	0.5A	20Ω	5W

Circuit B	V	I	R	P
R_1	6.7V	0.67A	10Ω	4.5W
R_2	3.3V	0.33A	10Ω	1.1W
R_3	3.3V	0.33A	10Ω	1.1W
Total	10V	0.67A	15Ω	6.7W

Circuit C	V	I	R	P
R_1	10V	1A	10Ω	10W
R_2	10V	1A	10Ω	10W
Total	10V	2A	5Ω	20W

Circuit D	V	I	R	P
R_1	8V	0.8A	10Ω	6.4W
R_2	4V	0.4A	10Ω	1.6W
R_3	4V	0.4A	10Ω	1.6W
R_4	8V	0.8A	10Ω	6.4W
Total	20V	0.8A	25Ω	16W

As circuits become more complex, with multiple series and parallel paths as well as multiple sources of potential difference, the VIRP table analysis you've been doing may become cumbersome. In these cases, you may find a direct application of Kirchhoff's Voltage Law and Kirchhoff's Current Law are more efficient methods of circuit analysis.

6.45 Q: A 50-ohm and 100-ohm resistor are connected as shown to a battery with emf of 40 volts and an internal resistance of r.

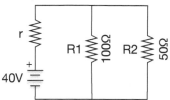

A) Find the value of r if the current in the circuit is 1 amp.

B) What is the battery's terminal voltage?

6.45 A: A) First combine resistors R1 and R2 to create an equivalent resistor R12, allowing you to solve for the internal resistance of the circuit. (Syntax Note: R_1 indicates the resistance of resistor R1.)

$$R_{12} = \frac{R_1 R_2}{R_1 + R_2} = \frac{(100\Omega)(50\Omega)}{100\Omega + 50\Omega} = 33\Omega$$

Redrawing the circuit in this simplified configuration you obtain:

Now you can determine the value of the internal resistance r by applying Kirchhoff's Voltage Law around the closed loop. Starting from the lower-left-hand corner of the circuit, you first see the negative side of the battery, so -40V, then the voltage drop across the internal resistance, Ir, then the voltage drop across R12, equal to $I \times R_{12}$. This brings you back to your starting point, so the sum of all these voltage drops must be zero:

$$-40V + Ir + I(33\Omega) = 0 \rightarrow (r + 33)I = 40 \xrightarrow{I=1A}$$
$$33 + r = 40 \rightarrow r = 7\Omega$$

B) Start by finding the voltage drop across the internal resistance using Ohm's Law:

$$V_r = Ir = (1A)(7\Omega) = 7V$$

The terminal voltage of the battery, then, is the battery's emf minus the voltage drop across the internal resistor.

$$V_T = \varepsilon - Ir = 40V - 7V = 33V$$

6.46 Q: A circuit with two voltage sources is depicted in the schematic at right.

A) Find the current flowing through R3.

B) Determine the power dissipated in R3.

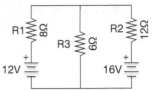

6.46 A: A) Start by applying Kirchhoff's Voltage Law (KVL) to the left-hand loop.

Beginning at the lower left and running clockwise, you first see the negative side of the battery (-12V), then the voltage drop across R1 ($8I_1$), then the voltage drop across R3 ($6I_3$), bringing you back to the original starting point, for a total voltage drop of zero.

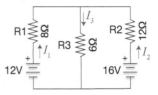

$$-12 + 8I_1 + 6I_3 = 0$$

Next, recognize using Kirchhoff's Current Law (KCL) that the sum of all the currents entering the top middle node of the circuit must equal the sum of all currents leaving. By defining I_1, I_2, and I_3 as shown in the diagram, you can amend your previous KVL equation as follows:

$$-12 + 8I_1 + 6I_3 = 0 \xrightarrow{I_3 = I_1 + I_2} -12 + 8I_1 + 6(I_1 + I_2) = 0 \rightarrow 14I_1 + 6I_2 = 12$$

Now, apply KVL to the right-hand loop using the same strategy. Beginning at the lower-right-hand corner and working counterclockwise, you first see the negative side of the battery (-16V), then the voltage drop across R2 ($12I_2$), then the voltage drop across R3 ($6I_3$), bringing you back to the original starting point.

$$-16 + 12I_2 + 6I_3 = 0 \xrightarrow{I_3 = I_1 + I_2} -16 + 12I_2 + 6(I_1 + I_2) = 0 \rightarrow 6I_1 + 18I_2 = 16$$

Combining these equations (two equations, two unknowns), you can solve for one of the unknowns:

$$-3(14I_1 + 6I_2) = -3(12) \rightarrow$$

$$-42I_1 - 18I_2 = -36$$

$$\underline{6I_1 + 18I_2 = 16}$$

$$-36I_1 \qquad = -20 \rightarrow$$

$$I_1 = 0.556A$$

With I_1 known you can now solve for I_2 and I_3.

$$6I_1 + 18I_2 = 16 \xrightarrow{I_1 = 0.556A} 6(0.556) + 18I_2 = 16 \rightarrow I_2 = 0.703A$$

$$I_1 + I_2 = I_3 \rightarrow I_3 = 1.26A$$

B) Find the power dissipated in R3 using $P = I^2 R$.

$$P_{R_3} = I^2 R = (1.26A)^2 (6\Omega) = 9.52W$$

6.47 Q: Given the schematic diagram below, determine the reading of both the ammeter and the voltmeter.

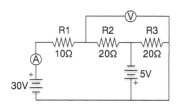

6.47 A: In examining the circuit, first recognize that the voltmeter draws a negligible current.

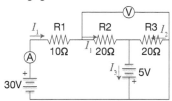

You can apply KVL around the bottom-left-hand-corner circuit loop in the clockwise direction by first noting the negative side of the battery (-30V), the voltage drop across R1 ($10I_1$), the voltage drop across R2 ($20I_1$), and the negative side of the 5V battery (-5V) before reaching your starting point, for a sum of potential drops of zero. From this, you can solve for I_1, which is the reading through the ammeter.

$$-30+10I_1+20I_1-5=0 \rightarrow 30I_1-35 \rightarrow I_1=1.17A$$

The reading across the voltmeter can be determined by applying KVL around the outside of the circuit, starting in the lower-left-hand-corner and moving clockwise. Beginning with the battery (-30V), the voltage drop across R1 ($10I_1$), and the voltage drop across the voltmeter (V), you return to the starting point, for a sum of potential drops of zero. From this equation, you can solve for the reading of the voltmeter V.

$$-30+10I_1+V=0 \rightarrow V=30-10I_1 \xrightarrow{I_1=1.17A} V=18.3V$$

6.48 Q: Identical light bulbs and batteries are employed in the following circuit configurations. Rank the brightness of the bulb in the various configurations from brightest to dimmest.

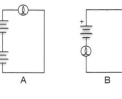

6.48 A: A=C, B=D=E. The brightness of the bulb is directly related to the power dissipated by the bulb. Since all the bulbs have the same resistance, an analysis of the potential drop across the bulbs leads you directly to the brightness, with greater potential drops providing greater brightness. Bulbs A and C each have a voltage drop of 2V across them, while bulbs B, D, and E all have a voltage drop of V across them.

6.49 Q: Four identical electrical light bulbs, labeled A through D, are connected to a source of constant potential difference in an unknown circuit configuration. When all bulbs are in place, the bulbs all have the same brightness. Based on the following information, draw the circuit diagram for the voltage source and the bulbs.

A) When bulb A is removed from the circuit, bulb C goes out. Bulbs B and D are unaffected.

B) When bulb B is removed from the circuit, bulb D goes out. Bulbs A and C are unaffected.

C) When bulb C is removed from the circuit, bulb A goes out. Bulbs B and D are unaffected.

C) When bulb D is removed from the circuit, bulb B goes out. Bulbs A and C are unaffected.

6.49 A: (Or equivalent)

Capacitors in Series and Parallel

Capacitors as electrical elements provide a variety of circuit functions. They can store energy and charge, they are used in various types of amplifiers, and they react in different ways to varying frequencies, making them useful in various signal processing applications. The symbol for a capacitor in a circuit is two parallel plates, as shown at right.

Capacitors in parallel can be replaced with an equivalent capacitor. You can find the capacitance of the equivalent capacitor by first recognizing that the voltage across the capacitors must be the same.

The charge on each capacitor is equal to the product of the capacitance and the voltage across it (Q=CV). You can then calculate the equivalent capacitance as follows:

$$C_{eq} = \frac{Q}{V} = \frac{Q_1 + Q_2}{V} = \frac{C_1 V + C_2 V}{V} = C_1 + C_2$$

This analysis can be generalized for any number of capacitors in parallel.

$$C_p = \sum_i C_i$$

In similar fashion, capacitors in series can be replaced with an equivalent capacitor.

The negative plate of Q_1 and the positive plate of Q_2 are electrically isolated, so their net charge must be zero. Therefore, each must have the same magnitude of charge. The alternate plates of capacitors must have equal magnitude of charge; therefore the charge on the plates of each of the capacitors in series must be equal, consistent with the law of conservation of electrical charge.

Start the analysis of the series capacitors by finding the potential difference across each of the capacitors, and adding them up to find the total potential difference across the capacitors.

$$C_{eq} = \frac{Q}{V_{total}} \rightarrow V_{total} = \frac{Q}{C_{eq}} = V_1 + V_2 \xrightarrow[V_2 = Q/C_2]{V_1 = Q/C_1} \frac{Q}{C_{eq}} = \frac{Q}{C_1} + \frac{Q}{C_2} \rightarrow \frac{1}{C_{eq}} = \frac{1}{C_1} + \frac{1}{C_2}$$

This relationship can be generalized for any numbers of capacitors in series:

$$\frac{1}{C_S} = \frac{1}{C_1} + \frac{1}{C_2} + \ldots = \sum_i \frac{1}{C_i}$$

Note how capacitors in parallel combine in a fashion similar to resistors in series, and capacitors in series combine in a fashion similar to resistors in parallel. Additionally, for configurations of two capacitors in series, the equivalent capacitance simplifies to:

$$C_S = \frac{C_1 C_2}{C_1 + C_2}$$

6.50 Q: Determine the equivalent capacitance of the capacitor network shown below.

6.50 A: First recognize that you can treat this as a configuration of three identical capacitors in series, all in parallel with a single capacitor, as re-drawn below.

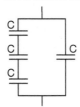

The equivalent capacitance of the three capacitors in series can then be determined as follows:

$$\frac{1}{C_S} = \sum_i \frac{1}{C_i} = \frac{1}{C_1} + \frac{1}{C_2} + \frac{1}{C_3} = \frac{1}{C} + \frac{1}{C} + \frac{1}{C} \rightarrow C_S = \frac{C}{3}$$

A simplified equivalent circuit can then be drawn as shown:

The equivalent capacitance is therefore just the parallel combination of the two remaining capacitors, which you find by taking the sum of their individual capacitances.

$$C_P = \sum_i C_i = \frac{C}{3} + C = \frac{4}{3}C$$

6.51 Q: What is the equivalent capacitance of the capacitor network shown below?

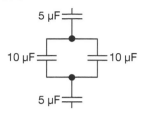

6.51 A: First combine the two 10 μF capacitors in parallel to obtain a 20 μF equivalent capacitance. This leads to a series network of capacitors, two 5-μF capacitors and a 20-μF capacitor.

$$\frac{1}{C_S} = \sum_i \frac{1}{C_i} = \frac{1}{5 \times 10^{-6}\,F} + \frac{1}{20 \times 10^{-6}\,F} + \frac{1}{5 \times 10^{-6}\,F} \rightarrow$$

$$C_S = 2.22 \times 10^{-6}\,F = 2.22\mu F$$

Capacitors in Circuits

Circuits containing a source of potential difference, a resistor network, and one or more capacitors are known as RC circuits. We'll explore RC circuits from the steady state perspective: what happens when they are first turned on, and what happens after a "long" time has elapsed. A detailed analysis of RC circuit operation as a function of time, known as a transient analysis, requires basic calculus, and therefore will be neglected in this treatment.

The key to understanding RC direct current circuit performance is remembering that uncharged capacitors act like wires, conducting current easily, while charged capacitors act like open circuits, blocking current flow.

Begin with a qualitative analysis of the charging RC circuit shown at right. At time t=0, the switch is closed and current begins to flow clockwise in the circuit. As it does so, positive charge accumulates on the top plate of the capacitor, along with a corresponding accumulation of negative charge on the bottom plate of the capacitor. This creates a potential difference across the capacitor, effectively reducing the potential difference across the resistor, and reducing the amount of current flow in the circuit. This continues until the voltage across the capacitor matches the terminal voltage of the battery, cutting the current flow of the circuit to zero as the capacitor reaches its maximum charge for the given circuit configuration.

A quantitative analysis of the circuit as it charges begins by applying Kirchhoff's Voltage Law to the loop. Beginning at the lower-left of the circuit and progressing clockwise, the sum of the voltage drops across each element leads to the following expression:

$$-V_T + IR + V_C = 0 \xrightarrow[V_C=Q/C]{C=Q/V_C} -V_T + IR + \frac{Q}{C} = 0 \xrightarrow{Q(t=0)=0}$$

$$-V_T + IR = 0 \rightarrow V_T = IR$$

This indicates that at time t=0, the circuit acts as if there were no capacitor (i.e. the capacitor acts like a wire).

After a long time, when the capacitor is fully charged, the capacitor acts like an open circuit (or "open"). The KVL analysis for this condition is considerably simpler, and validates that the final voltage across the capacitor is the voltage of the source.

$$-V_T + IR + V_C = 0 \xrightarrow{I=0} V_T = V_C$$

You can also look at the performance of the charging circuit from a graphical perspective, focusing on the current flow through the resistor, the charge on the capacitor, and the voltage across the plates of the capacitor.

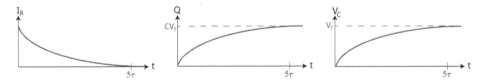

For each graph, the observed quantity follows an exponential relationship. The time constant τ (tau) in an RC circuit indicates the time at which the quantity under observation has achieved $(1-e^{-1})$, or 63%, of its final value. The observed quantities are within one percent of their final value in a time 5τ.

The RC time constant of the circuit is found by taking the product of the circuit's resistance and its capacitance (τ = RC), and is a useful tool for gauging how quickly a circuit responds to changes in state. Five time constants (5τ) is a useful approximation for the minimum time it takes for a circuit to reach its steady state condition.

Once a capacitor is charged, if the voltage source is removed, the capacitor can act like a source of potential difference (at least until it is discharged). Removing the voltage source from our previous RC circuit, you obtain the circuit shown at right.

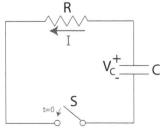

At time t=0, the switch is closed and the capacitor begins to discharge, acting as a source of potential difference while charge remains on its plates. As the charge is depleted, the potential difference across the plates of the capacitor decreases, reducing the potential difference driving the current in the circuit; therefore the current declines.

A quantitative analysis of the circuit as it discharges begins by applying Kirchhoff's Voltage Law to the loop. Beginning at the lower-right of the circuit and progressing counter-clockwise, the sum of the voltage drops across each element leads to the following expression:

$$-V_C + IR = 0 \rightarrow I = \frac{V_C}{R}$$

While there is charge on the plates of the capacitor, and therefore a potential across the plates, the capacitor acts like a voltage source and current flows through the circuit.

After a long time (approximately 5τ or longer), the capacitor is discharged and acts as a wire. Now KVL applied In the same manner leads to:

$$-V_C + IR = 0 \xrightarrow{V_C=0} I = 0$$

This indicates that once the capacitor has discharged, no current flows in the circuit.

Graphically, you can again focus on the current flow through the resistor, the charge on the plates of the capacitor, and the voltage across the plates of the capacitor.

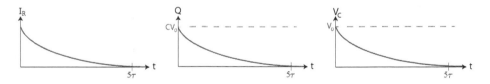

For each graph, again the observed quantity follows an exponential relationship related to the time constant, where the time constant τ is equal to RC. And again, the observed quantities are within one percent of their final value in a time 5τ.

6.52 Q: Given the circuit shown below, answer the following questions:
A) What is the current through R2 when the circuit is first connected?
B) What is the current through R2 a long time after the circuit has been connected?

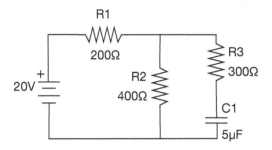

6.52 A: A) When the circuit is first connected, the capacitor is uncharged and therefore acts like a wire. The equivalent circuit could therefore be diagrammed as shown below.

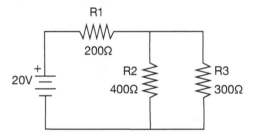

This could be simplified by recognizing that R2 and R3 are in parallel, and can be replaced by an equivalent resistor:

$$R_{eq} = \frac{R_2 \times R_3}{R_2 + R_3} = \frac{(400\Omega)(300\Omega)}{400\Omega + 300\Omega} = 171\Omega$$

The total current through the circuit can then be found using Ohm's Law:

$$I = \frac{V}{R} = \frac{20V}{(200\Omega + 171\Omega)} = 53.9mA$$

The voltage drop across both R2 and R3 can then be found using Ohm's Law:

$$V = IR = (0.0539A)(171\Omega) = 9.22V$$

Finally, solve for the current through R2 using Ohm's Law again:

$$I_2 = \frac{V_2}{R_2} = \frac{9.22V}{400\Omega} = 23.1mA$$

B) After the circuit has been connected a long time, the capacitor acts like an open. The equivalent circuit could therefore be diagrammed as shown below.

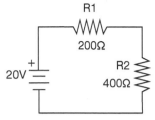

R1

200Ω

R2

400Ω

20V

As this is a series circuit, the current through R2 is the total current through the circuit, and can be easily obtained using Ohm's Law.

$$I = \frac{V}{R} = \frac{20V}{600\Omega} = 33.3mA$$

6.53 Q: The circuit below has been connected for a long time.

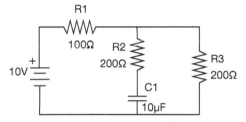

What is the current through R2 a long time after the circuit has been connected?

(A) 0 A

(B) 0.025 A

(C) 0.033 A

(D) 0.050 A

6.53 A: (A) 0 A. After the circuit has been connected for a long time, the capacitor acts like an open; therefore no current flows through R2.

6.54 Q: A circuit consisting of identical resistors and capacitors is shown below.

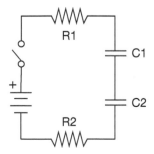

At time t=0, the switch is closed. Which of the following changes to the circuit are most likely to increase the amount of time it takes to charge the capacitors? Choose 2 correct answers.

(A) Adding an identical resistor in series with R2.

(B) Adding an identical resistor in parallel with R1.

(C) Adding a capacitor in series with C1.

(D) Adding a capacitor in parallel with C2.

6.54 A: (A) and (D) will both increase the RC time constant of the circuit, increasing the amount of time it takes to charge the capacitors.

6.55 Q: The following circuits are comprised of identical voltage sources, resistors, and capacitors. All switches are closed at the same time.

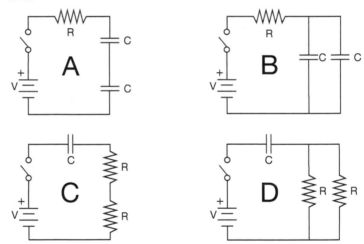

A) Rank the circuits from fastest to slowest in terms of how quickly they will reach their final steady-state condition.

B) Rank each circuit from highest to lowest current through the battery immediately after the switch is closed.

C) Rank each circuit from highest to lowest current through the battery a long time after the switch is closed.

D) Rank each circuit from largest to smallest amount of charge on the capacitor after a long time has elapsed.

E) Rank each circuit from largest to smallest potential difference across a single capacitor after a long time has elapsed.

6.55 A: A) A=D, B=C. Calculate the RC time constant of each circuit. The smallest RC time constants correspond to the fastest circuits.

B) D, A=B, C. The capacitors act like wires, to the current flow is inversely proportional to the resistance of the circuit.

C) A=B=C=D. The capacitors act like opens when fully charged, which in this case allows for no complete current paths, so the current in each circuit is zero.

D) B, C=D, A. The charge on the plates of the capacitor is the product of the equivalent capacitance and the source voltage, and since the source voltage is the same for all circuits, the charge is proportional to the equivalent capacitance.

E) B=C=D, A. After a long time, the current flow in each circuit is zero; therefore the entire potential difference across the source is dropped across the capacitors in circuits B, C, and D, while half the potential difference of the source is dropped across each of the capacitors in series in circuit A.

6.56 Q: In the circuit diagram below, capacitor C is fully charged at Q=15 µC. What is the value of the battery voltage?

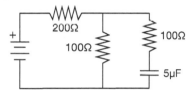

(A) 3 V

(B) 6 V

(C) 9 V

(D) 12 V

6.56 A: (C) 9 V. When fully charged, the voltage across the capacitor can be found from:

$$V = \frac{Q}{C} = \frac{15\mu C}{5\mu F} = 3V$$

Since the capacitor acts like an open, there is no current flowing through the right-most 100-ohm resistor; therefore the potential drop across the capacitor is the same as the potential drop across the remaining 100-ohm resistor. You can then use Ohm's Law to determine the current flow through the left-most 100-ohm resistor.

$$I = \frac{V}{R} = \frac{3V}{100\Omega} = 0.03A$$

This current must be the same as the current through the 200-ohm resistor, so you can find the voltage drop across the 200-ohm resistor using Ohm's Law.

$$V = IR = (0.03A)(200\Omega) = 6V$$

The battery voltage can then be found using Kirchhoff's Voltage Law around either of the loops going through the battery:

$$-V + 6V + 3V = 0 \rightarrow V = 9V$$

Test Your Understanding

1. What happens to the power usage in a room if you replace a 100-watt lightbulb with two 60-watt lightbulbs in parallel?

2. If all you have are 20-ohm resistors, how could you make a 5-ohm resistor? How could you make a 30-ohm resistor?

3. Explain how Kirchhoff's Current Law (the junction rule) is a restatement of the law of conservation of electrical charge.

4. Explain how Kirchhoff's Voltage Law (the loop rule) is a restatement of the law of conservation of energy.

5. What is the resistance of an ideal ammeter? What is the resistance of an ideal voltmeter? What would happen if you switched those meters in a circuit?

6. Draw a series circuit diagram consisting of three resistors and a battery on paper. From this circuit diagram, build a 3-D model of the circuit, where height is analogous to electric potential. Repeat for a parallel configuration.

7. Draw more complex circuit diagrams involving two or more voltage sources along with resistors. How would your 3-D models change with the additional voltage sources?

8. Electrical circuits are oftentimes compared to water flowing in pipes, where electrical current is analogous to water current, batteries are analogous to water pumps, and resistors are compared to water wheels. How could you extend this analogy to incorporate capacitors? Where do these analogies break down?

9. Graph the current through, voltage across, and power dissipated by a variable resistor attached to a constant-voltage battery as the resistor's resistance increases linearly with time.

10. Design an experiment to verify the marked capacitance of a capacitor.

11. Explain why the research and development of low-K dielectrics is such an important focus in the fabrication of microchips as more and more circuits are packed into smaller and smaller areas.

12. What happens to a string of Christmas lights if a single bulb goes out? What does this tell you about the circuit configuration?

Chapter 7: Magnetism

"Magnetism, as you recall from physics class,
is a powerful force that causes certain items
to be attracted to refrigerators."

— Dave Barry

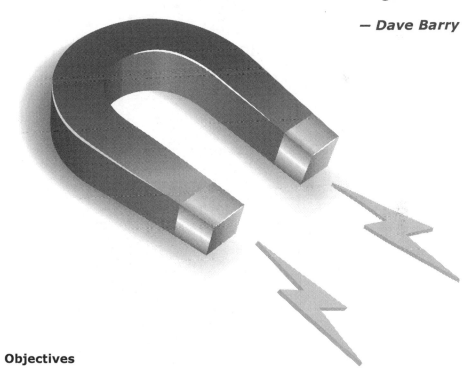

Objectives

1. Explain how magnetism is caused by moving electrical charges.
2. Describe the magnetic poles and interactions between magnets.
3. Describe magnetic properties such as magnetic permeability and magnetic dipole moment.
4. Draw magnetic field lines.
5. Calculate the force on a moving charge in a magnetic field.
6. Describe the magnetic field created by a current-carrying wire.
7. Describe the forces on current-carrying wires in magnetic fields.
8. Explain the concept of electromagnetic induction and apply fundamental principles to determine induced currents and emfs.

What Causes Magnetism?

Magnetism is closely related to electricity. In essence, **magnetism** is a force caused by moving charges. An electric field becomes a magnetic field when the electric charges are moving relative to the observer. But how does this lead to the creation of magnets and magnetism?

Moving charges create magnetic fields; therefore, all atoms have magnetic fields due to their moving electrons, known as orbital magnetic fields. Complete shells of electrons contain pairs of electrons which are considered to spin in opposite directions so that their magnetic fields cancel out, leaving no net magnetic field.

Materials comprised of atoms with filled outer electron shells are known as **diamagnetic** materials. In the presence of an external magnetic field, the electronic structure of diamagnetic materials tends to create a weak internal magnetic field opposing the external field.

Partially-filled shells of electrons, however, can give rise to a net magnetic field because the spinning electrons are unpaired. Therefore, atoms whose outermost electron shells are half-filled are the most magnetic atoms. But the story doesn't end there.

When atoms and molecules come together to make a solid, their constituent parts can be arranged in a variety of crystal lattice configurations. If all of the atoms or molecules align in the same direction to create a strong net magnetic field, the material is **ferromagnetic**. Ferromagnetic materials can be permanently magnetized.

On the other hand, it is possible that the atoms or molecules align themselves in random configurations, resulting in no net magnetic field, in which case the material is said to be **paramagnetic**. The magnetic orientation of paramagnetic materials can be weakly influenced by a magnetic field, but they do not remain aligned when the magnetic field is removed.

Further, a solid may be comprised of regions in which the magnetic fields of atoms are grouped and aligned, known as **magnetic domains**. If the magnetic domains are oriented in a random arrangement, there is no net magnetic field. If, however, the domains are aligned so that they point in the same direction, a net magnetic field is observed.

Random Domains

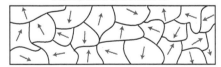

Aligned Domains

Magnetic domains may align naturally, or they can be aligned by application of an external magnetic field. You can demonstrate this by rubbing an iron nail repeatedly in the same direction with one end of a permanent magnet.

This causes the magnetic domains within the nail to align themselves in a common direction, converting the nail into a magnet.

In order to act as a strong permanent magnet, a material must be magnetic at three distinct configuration levels. At the atomic level, the atom must have a partially-filled outer shell. At the crystal lattice level, the atoms or molecules must be arranged in the same magnetic alignment. And at the domain level, all the magnetic domains must be oriented in the same direction.

As an additional complexity, at higher temperatures, the increased random thermal motion of the particles makes it more difficult for electrons to maintain their alignment, so higher temperatures typically reduce magnetic effects. In fact, only three elements are naturally magnetic at room temperature: iron, nickel, and cobalt. Gadolinium is paramagnetic at room temperature, but becomes ferromagnetic at temperatures just below room temperature (below about 20°C.)

Magnetic Fields

Magnets are polarized, meaning every magnet has two opposite ends. The end of a magnet that points toward the geographic north pole of the Earth is called the north pole of the magnet, while the opposite end, for obvious reasons, is called the magnet's south pole. Every magnet has both a north and a south pole. There are no single isolated magnetic poles, or monopoles. If you split a magnet in half, each half of the original magnet exhibits both a north and a south pole, giving you two magnets. Physicists continue to search both physically and theoretically, but to date, no one has ever observed a north pole without a south pole, or a south pole without a north pole.

You used electric field lines to help visualize what would happen to a positive charge placed in an electric field. In order to visualize a magnetic field, you can draw magnetic field lines (also known as magnetic flux lines) which show the direction the north pole of a magnet would tend to point if placed in the field. Magnetic field lines are drawn as closed loops, starting from the north pole of a magnet and continuing to the south pole of a magnet. Inside the magnet itself, the field lines run from the south pole to the north pole.

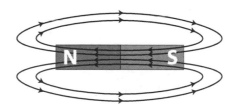

The magnetic field is strongest in areas of greatest density of magnetic field lines, or areas of the greatest magnetic flux density. Magnetic field strength (B) is measured in units known as Tesla (T).

$$1T = 1\frac{N \bullet s}{C \bullet m}$$

Much like electrical charges, like poles exert a repelling force on each other, while opposite poles exert an attractive force on each other.

When a magnetic field interacts with a material, the material obtains an amount of magnetization due to that field. The material's ability to support that magnetic field within itself is known as the material's **magnetic permeability** (μ). Materials with high magnetic permeability such as iron support stronger magnetic fields, while materials with lower magnetic permeability such as air and water support only weak magnetic fields. The permeability of a vacuum is an important constant in physics, given as $\mu_0 = 4\pi \times 10^{-7}$ (T$\bullet$m)/A.

7.01 Q: Which type of field is present near a moving electric charge?

(A) an electric field, only

(B) a magnetic field, only

(C) both an electric field and a magnetic field

(D) neither an electric field nor a magnetic field

7.01 A: (C) An electric field is present due to the electric charge, and a magnetic field is present because the charge is in motion.

7.02 Q: The diagram below shows the lines of magnetic force between two north magnetic poles. At which point is the magnetic field strength greatest?

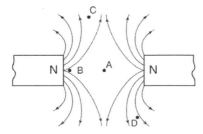

7.02 A: (B) has the greatest magnetic field strength because it is located at the highest density of magnetic field lines.

7.03 Q: The diagram below represents a 0.5-kilogram bar magnet and a 0.7-kilogram bar magnet with a distance of 0.2 meter between their centers.

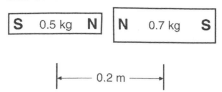

Which statement best describes the forces between the bar magnets?

(A) Gravitational force and magnetic force are both repulsive.

(B) Gravitational force is repulsive and magnetic force is attractive.

(C) Gravitational force is attractive and magnetic force is repulsive.

(D) Gravitational force and magnetic force are both attractive.

7.03 A: (C) Gravity always attracts and the north poles repel each other.

7.04 Q: A student is given two pieces of iron and told to determine if one or both of the pieces are magnets. First, the student touches an end of one piece to one end of the other. The two pieces of iron attract. Next, the student reverses one of the pieces and again touches the ends together. The two pieces attract again. What does the student definitely know about the initial magnetic properties of the two pieces of iron?

7.04 A: At least one of the pieces of iron is a magnet, but we cannot state with certainty that both are magnets.

7.05 Q: Draw a minimum of four field lines to show the magnitude and direction of the magnetic field in the region surrounding a bar magnet.

7.05 A:

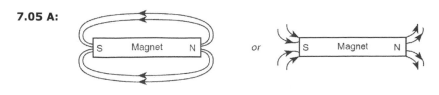

7.06 Q: When two ring magnets are placed on a pencil, magnet A remains suspended above magnet B, as shown below.

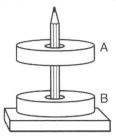

Which statement describes the gravitational force and the magnetic force acting on magnet A due to magnet B?

(A) The gravitational force is attractive and the magnetic force is repulsive.

(B) The gravitational force is repulsive and the magnetic force is attractive.

(C) Both the gravitational force and the magnetic force are attractive.

(D) Both the gravitational force and the magnetic force are repulsive.

7.06 A: (A) Gravity can only attract, and because magnet A is suspended above magnet B, the magnetic force must be repulsive.

The Compass

Because the Earth exerts a force on magnets (which, when used to tell direction, we call a compass), you can conclude that the Earth is a giant magnet. If the north pole of a magnet is attracted to the geographic north pole of the Earth, and opposite poles attract, then it stands to reason that the geographic north pole of the Earth is actually a magnetic south pole. Compasses always line up with the net magnetic field.

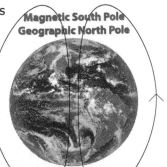

In truth, the magnetic north and south pole of the Earth are constantly moving. The current rate of change of the magnetic north pole is thought to be more than 20 kilometers per year, and it is believed that the magnetic north pole has shifted more than 1000 kilometers since it was first reached by explorer Sir John Ross in 1831!

7.07 Q: The diagram below represents the magnetic field near point P.

If a compass is placed at point P in the same plane as the magnetic field, which arrow represents the direction the north end of the compass needle will point?

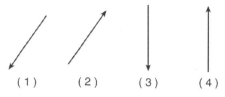

(1)   (2)   (3)   (4)

7.07 A: (2) Compass needles line up with the magnetic field.

7.08 Q: The diagram below shows a bar magnet.

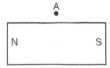

Which way will the needle of a compass placed at A point?

(A) up

(B) down

(C) right

(D) left

7.08 A: (C) since a compass lines up with the magnetic field.

Electromagnets*

In 1820, Danish physicist Hans Christian Oersted found that a current running through a wire created a magnetic field, kicking off the modern study of **electromagnetism**. You can test this by placing a compass near a current-carrying wire. The compass will line up with the induced magnetic field.

Building an understanding of why this occurs requires a brief foray into Einstein's Theory of Special Relativity. According to Special Relativity, length and time are not absolute measures, but rather they can be perceived differently based on the motion of the observers relative to each other. Specifically, moving objects contract in the direction of their motion relative to a stationary observer.

Applying this to current in a wire, first recognize that a wire contains a large number of positive metal ions fixed in space, surrounded by a "sea" of negative electrons which are able to move freely. The wire as a whole is neutral, however, because the quantity of positive ions matches the quantity of negative electrons.

When current flows through the wire, the net flow of negative electrons is in a specific direction, while the positive charges remain fixed in space. The density of positive charges and negative charges in any given section of wire is the same, therefore the wire as a whole is neutral. Note the equal number of positive and negative charges depicted in the section of wire below.

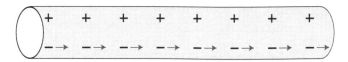

Now imagine a charged "observer object" moving outside the wire. As the observer object moves, the ions and electrons in the wire experience different motion relative to the observer object, so the separation of the ions and electrons differ slightly due to length contraction from the observer's perspective. This creates a difference in charge density between the positive ions and negative electrons in the wire from the observer object's perspective as shown in the diagram below, leading to a non-zero net electrical charge and therefore a net electric field.

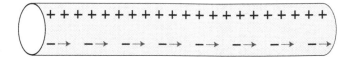

The charged observer object "sees" the wire as having a net electrical charge, and therefore it experiences an electric force from the wire. When this force is observed from a stationary frame of reference with respect to the moving charged object, it is known as a magnetic force. Ultimately, the magnetic force is just an electric force acting on a moving charged object.

Chapter 7: Magnetism

Forces on Moving Charges

Not only do moving charges create magnetic fields, but relative motion between charges and a magnetic field can produce a force. The magnitude of the force (F_B) on a charge (q) moving through a magnetic field (B) with a velocity (v) is given by:

$$F_B = qvB \sin \theta$$

In this equation, θ is the angle between the velocity vector and the direction of the magnetic field. If the velocity of the charged particle is perpendicular to the magnetic field, sin θ = sin 90° = 1, and the force can be calculated as simply F_B=qvB.

Because force is a vector, it has a direction as well. This direction can be determined using a right-hand rule. Point the fingers of your right hand in the direction of a positive particle's velocity (if the moving charge is negative, use your left hand). Then, curl your fingers inward 90° in the direction the magnetic field points. Your thumb will point in the direction of the force on the charged particle.

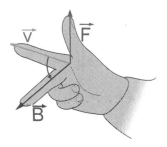

A scientific application of this principle involves the velocity selector, which acts as a filter to only allow charged particles of a specific velocity to pass through it. In this case, a uniform electric field is directed perpendicular to a uniform magnetic field, as shown in the diagram at right.

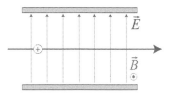

A positively charged particle moving to the right through the velocity selector is subjected to an upward electric force from the electric field and a downward magnetic force from the magnetic field. The strength of the electric and magnetic fields are chosen such that the magnitudes of the electric and magnetic forces are equal for a charged particle of the desired velocity. Charged particles with velocity other than the selected velocity are accelerated out of the linear path, while particles with the selected velocity feel no net force and continue through the device undeflected.

$$\left|\vec{F}_e\right|=\left|\vec{F}_B\right|\rightarrow qE=qvB\rightarrow v=\frac{E}{B}$$

Another application of this concept involves the mass spectrometer, a tool designed to measure the mass-to-charge ratio of individual molecules or atoms. It functions by ionizing a small sample of a material in a vacuum chamber, sorting the ions by their mass to charge ratio, and measuring the quantities of the sorted ions.

Once the particles are positively ionized, they are accelerated into a magnetic field, typically with a velocity selector. As the charged particles leave the velocity selector and enter the magnetic field, the magnetic field exerts a force on the particle perpendicular to its velocity, which is a centripetal force, causing the particle to turn in a circle (and doing no work). By measuring where the particle hits, you can determine the radius, and therefore calculate the mass of the particle.

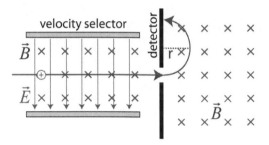

As the ion with charge q enters the magnetic field with some velocity v, the particle is accelerated in a circular path of radius r by the magnetic field (B) on the moving charge. The magnetic force provides the center-seeking, or centripetal, force. Combining the formulas for the magnitude of the magnetic force on a moving charged particle with the formula for the magnitude of the centripetal force, you can solve for the mass of the charged particle as follows:

$$F_C=\frac{mv^2}{r}=qvB\sin\theta\xrightarrow[\sin\theta=1]{\theta=90°}\frac{mv^2}{r}=qvB\rightarrow m=\frac{qrB}{v}$$

7.09 Q: The diagram below shows a proton moving with velocity v about to enter a uniform magnetic field directed into the page. As the proton moves in the magnetic field, the magnitude of the magnetic force on the proton is F.

X   X   X   X

(+) v→   X   X   X   X   Magnetic Field
                              Directed into
          X   X   X   X   the Page

X   X   X   X

If the proton were replaced by an alpha particle (charge +2e) under the same conditions, the magnitude of the magnetic force on the alpha particle would be

(A) F

(B) 2F

(C) F/2

(D) 4F

7.09A: (B) 2F. Because charge is doubled, magnetic force also doubles.

7.10 Q: An electron moves at 2×10^6 meters per second perpendicular to a magnetic field having a flux density of 2.0 teslas. What is the magnitude of the magnetic force on the electron?

7.10 A: $F_B = qvB\sin\theta = (-1.6\times10^{-19}C)(2\times10^6\,m/_s)(2T)\sin 90° = 6.4\times10^{-13}N$

7.11 Q: A mass spectrometer is designed such that the electric field strength in the velocity selector portion of the tool is **E**, and the magnetic field strength in both the velocity selector and the deflection region is **B**. What is the radius of the path for a singly charged positive ion of mass m?

(A) qE/mB²

(B) mE²/qB³

(C) qBE/m

(D) mE/qB²

7.11 A: (D) mE/qB²

The velocity of the incoming ions is determined by the velocity selector, where v=E/B. The radius of the path can be determined by setting the centripetal force equal to the magnetic force on the ion and solving for the radius.

$$F_c = F_B \rightarrow \frac{mv^2}{r} = qvB\sin\theta \xrightarrow{\sin 90°=1} r = \frac{mv^2}{qvB} \xrightarrow{v=E/B} r = \frac{mE}{qB^2}$$

7.12 Q: A particle of given charge and velocity is situated in a uniform magnetic field as shown in the following scenarios. Rank the magnitude of the magnetic force on the particles from greatest to least.

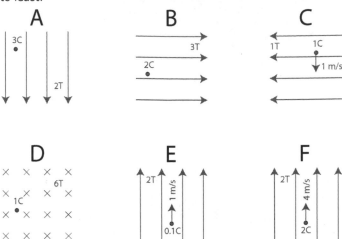

7.12 A: C, A=B=D=E=F=0

The magnetic force on the charges in all the scenarios is 0 except for C. In scenarios A, B, and D the charge isn't moving, so there is no magnetic force. In scenarios E and F, the charges move in the same direction as the magnetic field, so there is no magnetic force on the charge.

7.13 Q: In the scenarios below, a charged particle is deflected in a circular path as it enters a region of uniform magnetic field. For each scenario, determine the direction of the magnetic field.

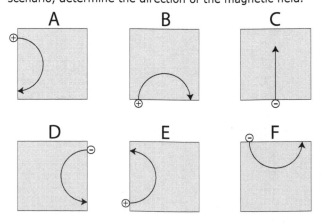

7.13 A: Using right-hand-rule:

   A) out of page
   B) out of page
   C) toward top or bottom of page
   D) out of page
   E) into page
   F) out of page

7.14 Q: A particle with a charge of 6.4×10⁻¹⁹ coulombs experiences a force of 2×10⁻¹² newtons as it travels through a three-tesla magnetic field at an angle of 30 degrees to the field. What is the particle's velocity?

7.14 A: $F_B = qvB\sin\theta \rightarrow v = \dfrac{F_B}{qB\sin\theta} \rightarrow$

$$v = \frac{2\times10^{-12}\,N}{(6.4\times10^{-19}\,C)(3T)(\sin30°)} = 2.08\times10^6\,{}^m\!/_s$$

Forces on Current-Carrying Wires

Magnetic fields cause a force on moving charges, and current-carrying wires contain moving charges; therefore a current-carrying wire in a magnetic field may experience a magnetic force. The magnitude of this force (F_B) can be found from the current in the wire (I), the length of the wire (*l*), the magnetic field strength (B), and the angle between the direction of current flow and the magnetic field (θ) as follows:

$$F_B = IlB\sin\theta$$

The direction of this force can be found using a right-hand rule. Point the fingers of your right hand in the direction a positive charge would flow (the direction of conventional current). Bend your fingers inward in the direction of the magnetic field. Your thumb will then point in the direction of the magnetic force on the wire. Note that this is the same direction as the force on each individual charge.

7.15 Q: A 10-meter wire carries 10 amperes of current through a five-tesla magnetic field directed into the plane of the page.

B

× × × × × × × × × × × ×

× × × × × × × × × × × ×

× × × × × × × × × × × × × I

━━━━━━━━━━━━━━━▶

× × × × × × × × × × × ×

× × × × × × × × × × × ×

× × × × × × × × × × × ×

Determine the magnitude and direction of the magnetic force on the wire.

7.15 A: The force is toward the top of the page (from the right hand rule.)

$$F_B = IlB\sin\theta = (10A)(10m)(5T) = 500N$$

7.16 Q: A U-shaped wire is placed in a tube and submerged in fresh water where there exists an 8T magnetic field as shown below.

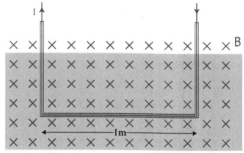

If the mass of the tube and wire is 1 kg, and the volume of the submerged tube is 0.003 m³, determine the current required to keep the system in static equilibrium.

7.16 A: First recognize that at equilibrium, the net force must be zero. A free-body diagram is a great way to start (be careful not to mix up the magnetic force and the buoyant force, both typically labeled F_B). Next, utilize Newton's 2nd Law in the y-direction, recognizing the acceleration in the y-direction is zero, to solve for the current through the wire.

$$F_{Buoyant} = mg + F_B \rightarrow \rho Vg = mg + IlB \rightarrow I = \frac{\rho Vg - mg}{lB}$$

$$I = \frac{(1000\,^{kg}\!/_{m^3})(0.003m^3)(10\,^{m}\!/_{s^2}) - (1kg)(10\,^{m}\!/_{s^2})}{(1m)(8T)} = 2.5A$$

7.17 Q: A loop of wire carrying current I is placed in a magnetic field. Determine the net torque around the axis of rotation due to the current in the wire at the instant the loop is placed.

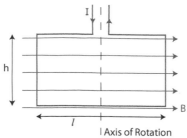

7.17 A: The only sections of current-carrying wire that run perpendicular to the magnetic field are the vertical sections, so by finding the force on each of those sections, you can find the torque. Note that the force on each side will be equal in magnitude but in opposite directions (out of the plane of the page on the left, into the plane of the page on the right), creating a net torque about the axis of rotation.

$$F_{left} = IlB\sin\theta = IhB = F_{right}$$

$$\tau_{left} = Fr\sin\theta = IhB\frac{l}{2} = \tau_{right}$$

$$\tau_{net} = \tau_{left} + \tau_{right} = IhBl$$

7.18 Q: A 5-meter long straight wire runs at a 45-degree angle to a uniform magnetic field of 5 tesla. If the force on the wire is 1 newton, determine the current in the wire.

7.18 A: $F_B = IlB\sin\theta \rightarrow I = \dfrac{F}{lB\sin\theta} \rightarrow I = \dfrac{1N}{(5m)(5T)\sin45°} = 0.0566A = 56.6mA$

Fields due to Current-Carrying Wires

Moving electric charges create magnetic fields; therefore electrical current in a wire creates a magnetic field around the wire. To determine the direction of the electrically-induced magnetic field, use a right hand rule (RHR) by pointing your right-hand thumb in the direction of positive current flow. The curve of your fingers as you grasp the wire shows the direction of the magnetic field around a wire.

Below left, the current is going into the page, creating a clockwise magnetic field. Below center, the current is coming out of the page, creating a counter-clockwise magnetic field. And below right, the current is flowing to the right of the page, creating a circular magnetic field which comes out of the page above the wire, and goes into the page below the wire.

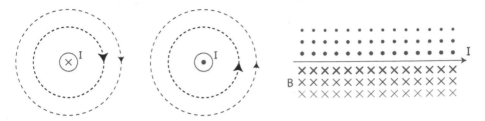

The magnetic field strength (B) is a function of the distance from the center of the wire (r). A stronger magnetic field exists close to the wire, and a weaker magnetic field exists further away from the wire, as described by the following relationship:

$$B = \frac{\mu_0}{2\pi} \frac{I}{r}$$

7.19 Q: A wire carries a current of 6 amperes to the left. Find the magnitude and direction of the magnetic field at point P, located 0.1 meter below the wire.

7.19 A: The direction of the magnetic field will be out of the plane of the page according to the right-hand rule. The magnetic field strength is found as follows:

$$B = \frac{\mu_0}{2\pi} \frac{I}{r} = \frac{4\pi \times 10^{-7} \, T \cdot m/A}{2\pi} \times \frac{6A}{0.1m} = 1.2 \times 10^{-5} T$$

7.20 Q: Two wires carry current as shown in the diagram below. Find the net magnetic field at point P.

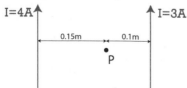

7.20 A: First find the strength of the magnetic field at point P due to the left-wire (into the plane of the page).

$$B_{left} = \frac{\mu_0}{2\pi}\frac{I}{r} = \frac{4\pi \times 10^{-7}\ ^{T\cdot m}\!/_A}{2\pi} \times \frac{4A}{0.15m} = 5.33 \times 10^{-6}T$$

Then find the strength of the magnetic field at point P due to the right-wire (out of the plane of the page).

$$B_{right} = \frac{\mu_0}{2\pi}\frac{I}{r} = \frac{4\pi \times 10^{-7}\ ^{T\cdot m}\!/_A}{2\pi} \times \frac{3A}{0.1m} = 6.0 \times 10^{-6}T$$

Find the net magnetic field strength by adding up the magnetic field due to each of the wires, recognizing that they are in opposite directions, and the net magnetic field will be out of the page.

$$B_P = B_{left} + B_{right} = 6.0 \times 10^{-6}T - 5.33 \times 10^{-6}T = 6.7 \times 10^{-7}T$$

7.21 Q: An electron moves parallel but opposite in direction to a current-carrying wire as shown in the diagram below. What is the direction of the force on the wire?

_____ I

(A) up
(B) down
(C) left
(D) right

7.21 A: (A) up. The wire creates a magnetic field coming out of the page above the wire. The magnetic force on the electron is down toward the bottom of the page by the right-hand-rule. Applying Newton's 3rd Law, the force on the wire must be up toward the top of the page.

You can obtain an even stronger magnetic field by wrapping a coil of wire in a series of loops known as a solenoid and flowing current through the wire. This is known as an electromagnet. You can make the magnetic field from the electromagnet even stronger by placing a piece of iron inside the coils of wire. Another right hand rule tells you the direction of the magnetic field due to an electromagnet. Wrap your fingers around the solenoid in the direction of positive current flow. Your thumb will point toward the north end of the induced magnetic field.

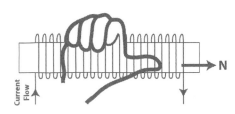

The strength of the magnetic field due to a solenoid (B_S) can be determined using the relationship below, where n is the number of loops of wire per unit length, and I is the current through the wire.

$$B_S = \mu_0 nI$$

7.22 Q: A solenoid of length 10 cm contains 150 coils of wire and carries a current of 10 mA. Determine the magnitude of the magnetic field created inside the solenoid.

7.22 A: $B_S = \mu_0 nI = (4\pi \times 10^{-7}\ {}^{T \cdot m}\!/\!_A) \left(\dfrac{150 coils}{0.1m} \right)(0.01A) = 1.88 \times 10^{-5} T$

7.23 Q: The air core of the electromagnet is replaced with an iron core. Compared to the strength of the magnetic field in the air core, the strength of the magnetic field in the iron core is

(A) less

(B) greater

(C) the same

7.23 A: (B) An iron core placed within an electromagnet strengthens the magnetic field.

Current-carrying wires create magnetic fields, and current-carrying wires contain moving charges which experience a force when moving through a magnetic field. Therefore, current-carrying parallel wires exert forces on each other.

Consider two current-carrying parallel wires as shown. The moving charges in the left wire create a magnetic field around the wire. The moving charges in the right wire feel a force due to the magnetic field from the left wire, resulting in a force on the right wire. In similar fashion, the right wire creates a magnetic field resulting in a force on the left wire. The forces on these parallel wires are equal in magnitude and opposite in direction, consistent with Newton's 3rd Law of Motion.

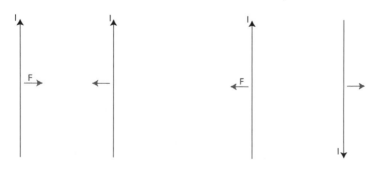

If the current in two parallel wires flows in the same direction, the wires are attracted to each other. If the current flows in opposite directions, the wires are repelled from each other.

Electromagnetic Induction

Much as charges flowing through a wire create magnetic fields, changing magnetic fields may move charges, known as "inducing" a current in a process known as electromagnetic induction. A potential difference is created by a changing magnetic field, known as the induced electro-motive force (emf). This potential difference can move charges in wire segments as well as circuits. Note, however, that the emf is not really a force, but is rather a potential difference.

The amount of magnetic field passing through an area is known as the magnetic flux (Φ_B). The units of magnetic flux are webers, where one weber is a tesla meter squared. The magnetic flux (Φ_B) through a given area (A) due to a magnetic field (B) can be determined using the relationship:

$$\Phi_B = BA\cos\theta$$

In this relationship, B is the magnitude of the magnetic field strength, A is the area through which the flux must pass, and angle θ is the angle between the magnetic field and the normal to the defined area.

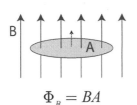

$$\Phi_B = BA$$

$$\Phi_B = 0$$

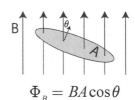

$$\Phi_B = BA\cos\theta$$

7.24 Q: Find the magnetic flux through a circular loop of wire of radius 0.2m sitting in a 3T uniform magnetic field if the loop of wire is tipped 20° from the direction of the magnetic field.

7.24 A: $\Phi_B = BA\cos\theta = (3T)(\pi \times (0.2m)^2)\cos 20° = 0.354W$

The magnitude of the induced emf is equal to the rate of change of the magnetic flux through an area defined by a loop of wire. In creating an emf, the area of the loop can change, the magnetic field strength can change, or the angle change. In all cases, however, the induced emf is given by:

$$\varepsilon = -\frac{\Delta\Phi_B}{\Delta t}$$

If the flux passes through multiple loops of wire, multiply the flux by the number of loops (N).

$$\varepsilon = -N\frac{\Delta\Phi_B}{\Delta t}$$

This phenomenon is what allows you to create usable, controllable electrical energy. Kinetic energy in the form of wind, water, steam, etc. is used to spin a coil of wire through a magnetic field, inducing a potential difference, which is transferred by the electric company to end users. This basic energy transformation is the underlying principle behind hydroelectric, nuclear, fossil fuel, and wind-powered electrical generators!

The direction of the current induced in the loop of wire always opposes the change in magnetic flux. This is known as **Lenz's Law**. The diagrams below illustrate this principle.

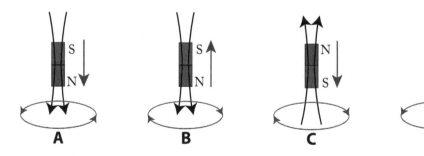

In diagram A, the magnetic flux down through the loop is increasing as the magnet is moved closer to the loop due to the increased density of magnetic lines of flux as you get closer to the pole of the magnet. The induced current opposes this change in flux, resulting in an induced magnetic field up out of the loop, corresponding to a counter-clockwise induced current in the loop. In diagram B, the magnetic flux down through the loop is decreasing as the magnet is moved away from the loop. The induced current in the wire opposes this change in flux by creating a magnetic flux down through the loop, corresponding to a clockwise current in the loop. Continuing examples are shown in diagrams C and D.

Chapter 7: Magnetism

By turning a coil of wire in a magnetic field, you can generate an induced emf. Generators use mechanical energy to turn the coil of wire, or to turn a magnet inside a coil of wire, creating a source of potential difference. Electric motors operate using the same principle, though in the case of the electric motor, electrical energy is the input and the output is the spinning wire (mechanical energy).

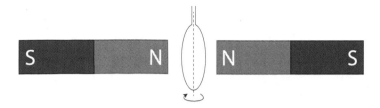

7.25 Q: A circular wire of radius 0.2m is sitting in a 3T uniform magnetic field. What is the induced emf in the wire if the hoop is rotated from 20° to 70° with respect to the magnetic field in five seconds? What is the direction of the Induced emf?

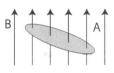

7.25 A: $\Phi_{B_i} = BA\cos\theta = (3T)(\pi \times (0.2m)^2)\cos 0° = 0.354W$

$\Phi_{B_f} = BA\cos\theta = (3T)(\pi \times (0.2m)^2)\cos 70° = 0.129W$

$\varepsilon = -\dfrac{\Delta\Phi_B}{\Delta t} = -\dfrac{\Phi_{B_f} - \Phi_{B_i}}{5s} = -\dfrac{(.129W - .354W)}{5s} = 0.045V$

Direction is counter-clockwise consistent with Lenz's Law.

7.26 Q: A coil of wire with 50 turns, each with a radius of 2.5 cm, is situated in a 2-tesla uniform magnetic field. After half a second, the magnetic field has dissipated completely. Determine the average emf induced in the wire.

7.26 A: $\Phi_{B_i} = BA\cos\theta = (2T)(\pi \times (0.025m)^2)\cos 0° = 0.00393W$

$\Phi_{B_f} = BA\cos\theta = 0$

$\varepsilon = -N\dfrac{\Delta\Phi_B}{\Delta t} = --N\dfrac{\Phi_{B_f} - \Phi_{B_i}}{5s} = -50\dfrac{(0W - 0.00393W)}{0.5s} = 0.393V$

A popular scenario involves the induction of an emf in an expanding or contracting rectangular loop of wire, as shown below.

Assume the loop of wire is placed in a constant magnetic field, and the right-hand section of wire which completes the loop moves to the right with constant speed v as indicated. The flux through the loop is increasing; therefore an emf is induced which creates a counter-clockwise current in the loop. The magnitude of the induced emf can be found as follows by recognizing that the x-dimension of the loop is the product of the wire's velocity and the time it has been in motion:

$$|\varepsilon| = \frac{\Delta\Phi_B}{\Delta t} = \frac{BA}{t} = \frac{BLvt}{t} = BLv$$

7.27 Q: The diagram below represents a wire conductor, RS, positioned perpendicular to a uniform magnetic field directed into the page.

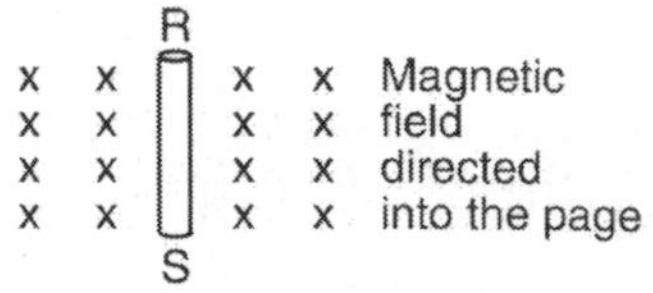

```
        R
 x  x  ┌─┐ x  x   Magnetic
 x  x  │ │ x  x   field
 x  x  │ │ x  x   directed
 x  x  └─┘ x  x   into the page
        S
```

Describe the direction in which the wire could be moved to produce the maximum potential difference across its ends, R and S.

7.27 A: The wire could be moved to produce the maximum potential difference across its ends, R and S, by moving it horizontally (right to left or left to right).

7.28 Q: The diagram below shows a wire moving to the right at speed v through a uniform magnetic field that is directed into the page.

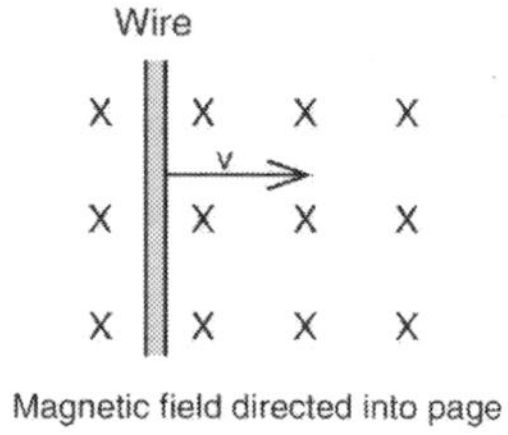

```
     Wire
  X ║ X   X   X
    ║   v→
  X ║ X   X   X
    ║
  X ║ X   X   X
```

Magnetic field directed into page

As the speed of the wire is increased, the induced potential difference will

(A) decrease

(B) increase

(C) remain the same

7.28 A: (B) the induced potential difference will increase as the speed of the wire is increased.

7.29 Q: A U-shaped loop of wire is connected by a conducting path on rails which moves upward at a constant 3 m/s.

If this wire is situated perpendicular to a uniform 0.5-tesla magnetic field, find the induced emf in the loop of wire. In which direction will the current flow?

7.29 A: $|\varepsilon| = Blv = (0.5T)(0.5m)(3^m/_s) = 0.75V$

Direction is counter-clockwise consistent with Lenz's Law

7.30 Q: The scenarios below depict a current-carrying wire in a region of uniform magnetic field. In all cases, the current and the length of the wire in the magnetic field are the same. Rank the magnitude of the magnetic force on the wire from greatest to least.

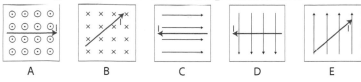

7.30 A: A=B=D, E, C=0.

7.31 Q: Six loops of wire of equal area are moving with speed v in a uniform magnetic field as shown in the diagram below. Rank the magnitude of the induced current in each of the loops at the point shown from greatest to least.

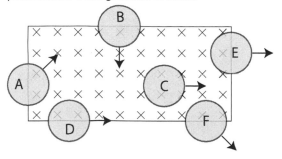

7.31 A: B=E, A=F, D=C=0.

Since all loops are of equal area and at the same angle, perpendicular to the magnetic field, the greatest induced current will occur for the greatest changes in magnetic flux.

7.32 Q: Two long straight current-carrying wires are situated near a point P as shown below. Rank the magnitude of the magnitude of the magnetic field at point P from greatest to least.

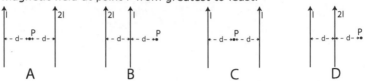

7.32 A: D, B, A, C=0

7.33 Q: A large wire sits above a compass as shown. What is the direction of the compass needle when a large current flows through the wire?

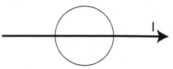

(A) toward top of page

(B) toward bottom of page

(C) toward the right

(D) toward the left

7.33 A: (A) toward top of page by right hand rule.

7.34 Q: A rectangular loop of wire attached to a resistor in series is pulled to the right at constant speed v in a region of constant magnetic field of strength B as shown. Determine the magnitude and direction of the induced current in the wire.

7.34 A: The induced emf can be found from:

$$\varepsilon = BLv$$

Using Ohm's Law, you can then find the current in the circuit as:

$$I = \frac{\varepsilon}{R} = \frac{BLv}{R}$$

The direction is clockwise as the induced current opposes the change in the magnetic flux.

7.35 Q: Two long parallel wires carrying current I are arranged a distance d apart as shown in the diagram, resulting in a force F on the right-most wire.

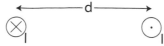

A) What is the direction of the net force on the right-most wire?

B) If the current in each wire is doubled and the distance between the wires is halved, what is the force on the right-most wire?

7.35 A: A) Right. The wires repel, so the force on the right-most wire is to the right.

B) 8F. Doubling the current in each wire doubles the force, so the force is four times larger due to the current change. Halving the distance between the wires doubles the force, therefore the total change in force is 8 times larger than the original force, or 8F.

7.36 Q: A square loop of wire of area A is situated in a constant magnetic field of strength B, with the left side of the loop attached to a fixed axis of rotation. What is the magnitude of the net torque on the loop if the loop carries a current I as shown?

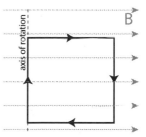

(A) BIA

(B) BI²A

(C) BIA²

(D) BI√A

7.36 A: (A) The net torque on the loop is Fl, where F is BIl, with l defined as the length of any given side of the loop. Combining these, the net torque is BIl^2, but l^2=A, so the net torque is BIA. Note that only the right-most section of wire contributed to the net torque on the loop.

7.37 Q: A circular wire loop sits in a magnetic field directed into the plane of the page as shown. Which of the following will induce a clockwise current in the loop? Select two answers.

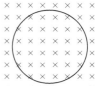

(A) increasing the size of the loop

(B) decreasing the size of the loop

(C) increase the strength of the magnetic field

(D) decreasing the strength of the magnetic field

7.37 A: (B) and (D) will result in clockwise currents consistent with Lenz's Law.

7.38 Q: A rectangular loop of wire situated on a frictionless cart is pulled by a rope through a constant magnetic field as shown below.

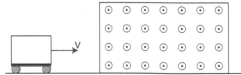

Assuming positive currents are clockwise and t=0 corresponds to the moment the front edge of the loop first enters the magnetic field, which graph best represents the induced current in the loop as a function of time?

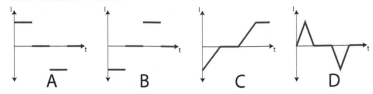

7.38 A: (A) As the loop first enters the magnetic field, the magnetic flux through the loop changes at a constant rate. Lenz's Law states that this will be a clockwise current, which is positive by the convention stated in the problem. Once the loop is completely in the magnetic field, there is no change in magnetic flux, so there is no current. Then, as the loop leaves the magnetic field, the magnetic flux through the loop decreases at a constant rate, inducing a counter-clockwise current in the loop. Once the loop has completely exited the magnetic field, there is no change in magnetic flux, so there is no induced current.

7.39 Q: A power supply is used to force a constant electrical current of 2 amperes through a metal Slinky toy (solenoid) with 50 turns of wire in the circuit as shown below.

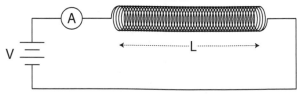

A magnetic field sensor is used to measure the strength of the magnetic field inside the Slinky's coils as the Slinky's length (L) is varied. The data from this experiment is shown below.

Length of Slinky (m)	Turns / Meter n	B Field Strength (T)
0.5		2.4×10^{-4}
1.0	50	1.3×10^{-4}
1.5		8.1×10^{-5}
2.0		6.5×10^{-5}

A) Complete the data table and plot the magnetic field strength on the y-axis vs. the turns per meter of the solenoid on the x-axis. Be sure to label all axes appropriately.

B) Determine the slope of the line of best fit.

C) Calculate the permeability of free space from the slope of your best-fit line.

D) Determine a percent error from the accepted value for the permeability of a vacuum.

7.39 A: A)

Length of Slinky (m)	Turns / Meter n (turns/m)	B Field Strength (T)
0.5	100	2.4×10^{-4}
1.0	50	1.3×10^{-4}
1.5	33.3	8.1×10^{-5}
2.0	25	6.5×10^{-5}

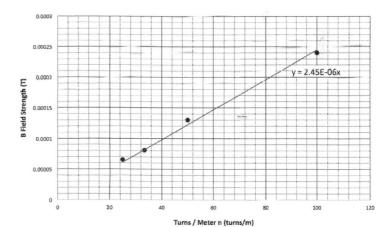

B) Slope = 2.45×10^{-6} T•m

C) B=$(\mu_0 I)n$, so the slope = $\mu_0 I$, therefore μ_0=slope/I= 2.45×10^{-6} T•m/(2A) = 1.23×10^{-6} T•m/A

D) % error $= \dfrac{|Approximate\,Value - Exact\,Value|}{|Exact\,Value|} \times 100\% \rightarrow$

% error $= \dfrac{\left|1.23 \times 10^{-6}\,{}^{T \bullet m}\!/_A - 4\pi \times 10^{-7}\,{}^{T \bullet m}\!/_A\right|}{\left|4\pi \times 10^{-7}\,{}^{T \bullet m}\!/_A\right|} = 2.1\%$

Test Your Understanding

1. Compare and contrast the magnetic force with the electrical force and the gravitational force.

2. In your own words, explain the relationship between electricity and magnetism.

3. In paragraph form, explain why iron can make a powerful permanent magnet, but an apple cannot.

4. Research how magnetic resonance imaging works.

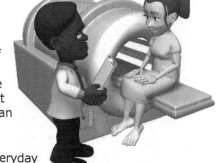

5. Build an electric motor from two magnets, two thumbtacks, a block of wood, two paperclips, insulated wire, and a flashlight battery. There are a variety of Internet resources that can help you. See how fast you can make your motor spin!

6. What are the uses of magnetism in everyday life?

7. Why do magnets attract and repel?

8. What causes Earth's magnetic field?

9. In problem 7.39, you neglected any effects of Earth's magnetic field. How would this experiment change if Earth's magnetic field were not neglected? Explain.

10. How is the aurora borealis related to magnetism?

11. Compare and contrast the workings of a traditional spinning hard drive in a computer with a solid state drive.

12. Explore the functioning of a hydroelectric power plant. Trace the energy conversions and see if you can make a complete loop in your energy trace.

13. How can electromagnetism be used in braking systems? Where is this used? Where is it not used? Why?

Chapter 8: Optics

"Music is the arithmetic of sounds as optics is the geometry of light."

— *Claude Debussy*

Objectives

1. Describe the creation and characteristics of electromagnetic waves.
2. Indicate how the polarization of a wave may affect its propagation.
3. Analyze the interference of waves using the superposition principle, including the creation of standing waves.
4. Predict frequency shifts in waves due to the Doppler Effect.
5. Analyze the behavior of concave and convex mirrors using the law of reflection and basic principles of geometric optics.
6. Analyze the behavior of concave and convex lenses using Snell's Law and basic principles of geometric optics.
7. Explain the diffraction of light in terms of Huygens' Principle.
8. Analyze and predict interference patterns for monochromatic light passing through single and double slits, as well as diffraction gratings.
9. Predict the behavior of light reflecting off of layers of thin films and interfering with itself.

Waves transfer energy through matter or space, and are found everywhere: sound waves, light waves, microwaves, radio waves, water waves, earthquake waves, Slinky waves, x-rays, and on and on. Light itself is an electromagnetic wave; therefore developing an understanding of waves in general as well as light and electromagnetic waves in particular will allow you to understand how energy is transferred in the universe, and will eventually lead to a better understanding of matter and energy itself!

Wave Characteristics

A **pulse** is a single disturbance which carries energy through a medium or through space. Imagine you and your friend holding opposite ends of a Slinky. If you quickly move your arm up and down, a single pulse will travel down the Slinky toward your friend.

If, instead, you generate several pulses at regular time intervals, you now have a wave carrying energy down the Slinky. A **wave**, therefore, is a repeated disturbance which carries energy. The mass of the Slinky doesn't move from one end of the Slinky to the other, but the energy it carries does.

When a pulse or wave reaches a hard boundary, it reflects off the boundary, and is inverted. If a pulse or wave reaches a soft, flexible boundary, it still reflects off the boundary, but does not invert.

Waves can be classified in several different ways. One type of wave, known as a **mechanical wave**, requires a medium, or material, through which to travel. Examples of mechanical waves include water waves, sound waves, Slinky waves, and even seismic waves. **Electromagnetic waves**, on the other hand, do not require a medium in order to travel. Electromagnetic waves (or EM waves) are considered part of the Electromagnetic Spectrum. Examples of EM waves include light, radio waves, microwaves, and even X-rays.

Further, waves can be classified based upon their direction of vibration. Waves in which the "particles" of the wave vibrate in the same direction as the wave velocity are known as **longitudinal**, or compressional, waves. Examples of longitudinal waves include sound waves and seismic P waves. Waves in which the particles of the wave vibrate perpendicular to the wave's direction of motion are known as **transverse** waves. Examples of transverse waves include seismic S waves, electromagnetic waves, and even stadium waves (the "human" waves you see at baseball and football games!).

Video animations of pulses reflecting off boundaries as well as longitudinal and transverse pulses can be viewed at http://bit.ly/gC1TMU.

Waves have a number of characteristics which define their behavior. Looking at a transverse wave, you can identify specific locations on the wave. The highest points on the wave are known as **crests**. The lowest points on the wave are known as **troughs**. The **amplitude** of the wave, corresponding to the energy of the wave, is the distance from the baseline to a crest or the baseline to a trough.

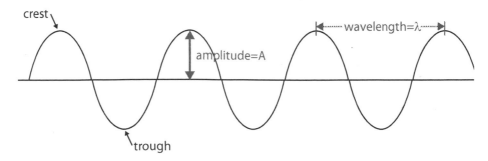

The length of the wave, or **wavelength**, represented by the Greek letter lambda (λ), is the distance between corresponding points on consecutive waves (i.e. crest to crest or trough to trough). Points on the same wave with the same displacement from equilibrium moving in the same direction (such as a crest to a crest or a trough to a trough) are said to be in phase (phase difference is 0° or 360°). Points with opposite displacements from equilibrium (such as a crest to a trough) are said to be 180° out of phase.

Simple waves can be described by basic sine and cosine functions, as described previously in our study of simple harmonic motion from AP Physics 1.

8.01 Q: Which type of wave requires a material medium through which to travel?
(A) sound
(B) television
(C) radio
(D) x ray

8.01 A: (A) sound is a mechanical wave and therefore requires a medium.

8.02 Q: The diagram below represents a transverse wave traveling to the right through a medium. Point A represents a particle of the medium.

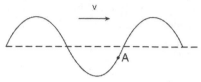

In which direction will particle A move in the next instant of time?

(A) up

(B) down

(C) left

(D) right

8.02 A: (B) particle A will move down as the wave passes.

8.03 Q: As a transverse wave travels through a medium, the individual particles of the medium move

(A) perpendicular to the direction of wave travel

(B) parallel to the direction of wave travel

(C) in circles

(D) in ellipses

8.03 A: (A) perpendicular to the direction of wave travel.

8.04 Q: A ringing bell is located in a chamber. When the air is removed from the chamber, why can the bell be seen vibrating but not be heard?

(A) Light waves can travel through a vacuum, but sound waves cannot.

(B) Sound waves have greater amplitude than light waves.

(C) Light waves travel slower than sound waves.

(D) Sound waves have higher frequency than light waves.

8.04 A: (A) Light is an EM wave, while sound is a mechanical wave.

8.05 Q: A single vibratory disturbance moving through a medium is called
(A) a node
(B) an antinode
(C) a standing wave
(D) a pulse

8.05 A: (D) a pulse.

8.06 Q: A periodic wave transfers
(A) energy only
(B) mass only
(C) both energy and mass
(D) neither energy nor mass

8.06 A: (A) energy only.

8.07 Q: The diagram below represents a transverse wave.

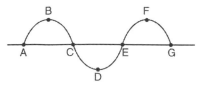

The wavelength of the wave is equal to the distance between points
(A) A and G
(B) B and F
(C) C and E
(D) D and F

8.07 A: (B) B and F is the wavelength as measured from crest to crest.

8.08 Q: The diagram below represents a periodic wave.

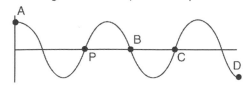

Which point on the wave is in phase with point P?

8.08 A: (C) Point C is the same point as point P but on a consecutive wave. Point B doesn't qualify because the wave has not completed a complete cycle.

8.09 Q: The diagram below represents a transverse wave moving on a uniform rope with point A labeled. On the diagram, mark an X at the point on the wave that is 180° out of phase with point A.

8.09 A:

8.10 Q: The diagram below represents a transverse wave traveling in a string.

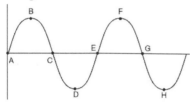

Which two labeled points are 180° out of phase?

(A) Points A and D

(B) Points B and F

(C) Points D and F

(D) Points D and H

8.10 A: (C) Points D and F.

8.11 Q: A periodic wave is produced by a vibrating tuning fork. The amplitude of the wave would be greater if the tuning fork were

(A) struck more softly

(B) struck harder

(C) replaced by a lower frequency tuning fork

(D) replaced by a higher frequency tuning fork

8.11 A: (B) Striking the tuning fork harder gives the tuning fork more energy, increasing the sound wave's amplitude.

8.12 Q: Increasing the amplitude of a sound wave produces a sound with

(A) lower speed

(B) higher pitch

(C) shorter wavelength

(D) greater loudness

8.12 A: (D) greater loudness due to the greater energy / amplitude of the wave.

The Wave Equation

The **frequency** (f) of a wave describes the number of waves that pass a given point in a time period of one second. The higher the frequency, the more waves that pass. Frequency is measured in number of waves per second (1/s), also known as a Hertz (Hz). If 60 waves pass a given point in a second, the frequency of the wave would be 60 Hz.

Closely related to frequency, the **period** (T) of a wave describes how long it takes for a single wave to pass a given point and can be found as the reciprocal of the frequency. Period is a measurement of time, and therefore is measured in seconds.

8.13 Q: What is the period of a 60-hertz electromagnetic wave traveling at 3.0×10^8 meters per second?

8.13 A: $T = \dfrac{1}{f} = \dfrac{1}{60 Hz} = 0.0167 s$

8.14 Q: Which unit is equivalent to meters per second?

(A) Hz•s

(B) Hz•m

(C) s/Hz

(D) m/Hz

8.14 A: (B) $\dfrac{m}{s} = Hz \bullet m$

8.15 Q: The product of a wave's frequency and its period is
(A) one
(B) its velocity
(C) its wavelength
(D) Planck's constant

8.15 A: (A) $f \bullet T = f \bullet \dfrac{1}{f} = 1$

Because waves move through space, they must have velocity. The velocity of a wave is a function of the type of wave and the medium it travels through. Electromagnetic waves moving through a vacuum, for instance, travel at roughly 3×10^8 m/s. This value is so famous and important in physics it is given its own symbol, **c**. When an electromagnetic wave enters a different medium, such as glass, it slows down. If the same wave were to then re-emerge from glass back into a vacuum, it would again travel at c, or 3×10^8 m/s.

The speed of a wave is determined by the wave type and the medium it is traveling through, and can be easily related to its frequency and wavelength. For a given wave speed, waves with higher frequencies have shorter wavelengths, and waves with lower frequencies have longer wavelengths. This can be shown mathematically using the wave equation.

$$v = f\lambda$$

8.16 Q: A periodic wave having a frequency of 5 hertz and a speed of 10 meters per second has a wavelength of
(A) 0.50 m
(B) 2.0 m
(C) 5.0 m
(D) 50. m

8.16 A: (B) $v = f\lambda$

$$\lambda = \frac{v}{f} = \frac{10\,m/s}{5Hz} = 2m$$

8.17 Q: If the amplitude of a wave is increased, the frequency of the wave will

(A) decrease
(B) increase
(C) remain the same

8.17 A: (C) remain the same.

8.18 Q: An electromagnetic wave traveling through a vacuum has a wavelength of 1.5×10^{-1} meters. What is the period of this electromagnetic wave?

(A) 5.0×10^{-10} s
(B) 1.5×10^{-1} s
(C) 4.5×10^{7} s
(D) 2.0×10^{9} s

8.18 A: (A) $v = f\lambda = \dfrac{\lambda}{T}$

$$T = \frac{\lambda}{v} = \frac{1.5 \times 10^{-1}\,m}{3 \times 10^{8}\,m/s} = 5 \times 10^{-10}\,s$$

8.19 Q: A surfacing blue whale produces water wave crests having an amplitude of 1.2 meters every 0.40 seconds. If the water wave travels at 4.5 meters per second, the wavelength of the wave is

(A) 1.8 m
(B) 2.4 m
(C) 3.0 m
(D) 11 m

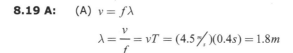

8.19 A: (A) $v = f\lambda$

$$\lambda = \frac{v}{f} = vT = (4.5\,m/s)(0.4s) = 1.8m$$

Electromagnetic Spectrum

Unlike mechanical waves, electromagnetic (EM) waves do not require a medium in which to travel. They consist of an electric field component and a magnetic field component oriented perpendicular to each other and to the wave velocity, and are caused by accelerating electrical charges. The orientation of the electric field and magnetic field components of an electromagnetic wave can be visualized below.

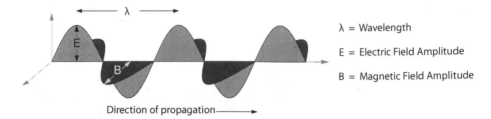

λ = Wavelength

E = Electric Field Amplitude

B = Magnetic Field Amplitude

Direction of propagation

Repeating from earlier, the speed of all electromagnetic waves in a vacuum is approximately 3×10^8 m/s, also known as c. This is, according to our current understanding of the universe, the fastest possible speed anything in the universe can travel.

Since c is a constant for all EM waves in a vacuum, the product of frequency and wavelength must be a constant. Therefore, at higher frequencies, EM waves have a shorter wavelength, and at lower frequencies, EM waves have a longer wavelength. If the EM wave travels into a new medium, its speed can decrease, and because frequency depends on the source and therefore remains constant, its wavelength would also decrease.

It is the frequency of an EM wave that determines its characteristics. The relationship between frequency and wavelength in a vacuum for various types of EM waves is depicted in the Electromagnetic Spectrum. This diagram can be useful for answering questions and solving problems involving electromagnetic waves.

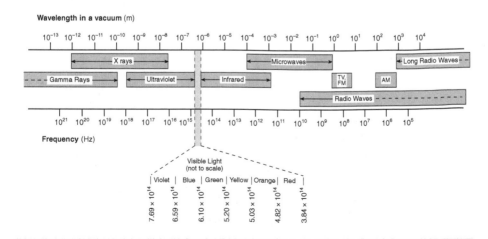

Chapter 8: Optics

The electromagnetic spectrum describes the types of electromagnetic waves observed at the specified frequencies and wavelength. It is also important to note that the energy of an electromagnetic wave is directly related to its frequency. Therefore, higher frequency (shorter wavelength) EM waves have more energy than lower frequency (longer wavelength) EM waves.

An x-ray, therefore, has considerably more energy than an AM radio wave! Using the diagram, more energetic waves are shown on the left side of the EM Spectrum, and less energetic waves are shown to the right on the EM Spectrum. You'll explore the energy of EM radiation further in the Modern Physics chapter.

8.20 Q: Which color of light has a wavelength of 5.0×10^{-7} meters in air?

(A) blue

(B) green

(C) orange

(D) violet

8.20 A: (B) First find the frequency using v=fλ, then use the Electromagnetic Spectrum to determine the correct color based on the frequency.

$$v = f\lambda$$

$$f = \frac{v}{\lambda} = \frac{c}{\lambda} = \frac{3\times10^{8}\,{}^{m}\!/\!{}_{s}}{5.0\times10^{-7}\,m} = 6\times10^{14}\,Hz$$

8.21 Q: A television remote control is used to direct pulses of electromagnetic radiation to a receiver on a television. This communication from the remote control to the television illustrates that electromagnetic radiation

(A) is a longitudinal wave

(B) possesses energy inversely proportional to its frequency

(C) diffracts and accelerates in air

(D) transfers energy without transferring mass

8.21 A: (D) transfers energy without transferring mass.

8.22 Q: A microwave and an x ray are traveling in a vacuum. Compared to the wavelength and period of the microwave, the x ray has a wavelength that is

(A) longer and a period that is shorter

(B) longer and a period that is longer

(C) shorter and a period that is longer

(D) shorter and a period that is shorter

8.22 A: (D) shorter and a period that is shorter.

8.23 Q: A 1.50×10⁻⁶-meter-long segment of an electromagnetic wave having a frequency of 6×10¹⁴ hertz is represented below.

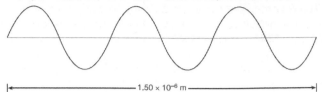

A) Mark two points on the wave that are in phase with each other. Label each point with the letter P.

B) Which type of electromagnetic wave does the segment in the diagram represent?

8.23 A: A)

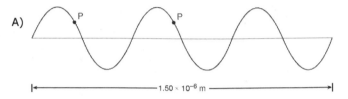

B) green light (visible light).

8.24 Q: What is the period of a 60-hertz electromagnetic wave traveling at 3.0×10⁸ meters per second?

(A) 1.7×10⁻² s

(B) 2.0×10⁻⁷ s

(C) 6.0×10¹ s

(D) 5.0×10⁶ s

8.24 A: (A) $T = \dfrac{1}{f} = \dfrac{1}{60 Hz} = 0.017 s$

Polarization

As mentioned previously, the directions of vibration of the electric field, magnetic field, and the velocity of an electromagnetic wave are all perpendicular to each other. Waves in which the electric and magnetic fields vibrate in more than one plane are called unpolarized, and are created by electric charges accelerating in a variety of directions. Waves in which the electric and magnetic fields vibrate in a single plane are polarized. A visual representation of the electric field of polarized and unpolarized light as it emerges from the page is shown below.

Certain materials made of long molecules arranged parallel to each other can permit certain orientations of EM waves to pass while blocking other orientations. This operation is similar to that of a picket fence. When unpolarized light passed through a **polarizing filter**, half the orientations are blocked, therefore half the intensity of the light is blocked as well. Two polarizing filters at right angles to each other are therefore opaque, blocking 100% of the incident light. The intensity of polarized light transmitted by a polarizing filter is given by **Malus' Law**, which states that $I = I_0 \cos^2\theta$, where I is the intensity of the incoming light and theta is the angle between the axis of the polarizing filter and the incoming light's polarization angle.

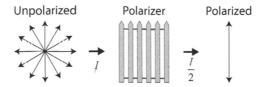

Polarizing filters are used in a variety of applications, from liquid crystal displays (LCDs) to sunglasses. Light from the sun is unpolarized, yet when it is reflected from a nonmetallic surface, it tends to orient itself horizontally. Polarized sunglasses include a polarizing filter oriented vertically, which blocks a significant amount of the reflected horizontal light, reducing glare from surfaces such as lakes, roads, and windshields.

8.25 Q: Unpolarized light of intensity I is incident upon a polarizing filter, leaving the filter as polarized light with intensity I/2. A second polarizing filter is added to the system, such that the intensity of the light leaving the filter system is I/4. What is the orientation of the 2nd polarizing filter with respect to the first?

8.25 A: The second filter must be situated at a 45° angle to the first filter to block half of the orientations of the polarized light.

Interference

When more than one wave travels through the same location in the same medium at the same time, the total displacement of the medium is governed by the principle of **superposition**. The principle of superposition simply states that the total displacement is the sum of all the individual displacements of the waves. The combined effect of the interaction of the multiple waves is known as **wave interference**.

8.26 Q: The diagram at right shows two pulses approaching each other in a uniform medium. Diagram the superposition of the two pulses.

8.26 A:

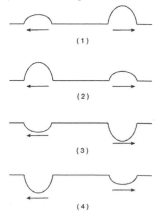

When two or more pulses with displacements in the same direction interact, the effect is known as **constructive interference**. The resulting displacement is greater than the original individual pulses. Once the pulses have passed by each other, they continue along their original path in their original shape, as if they had never met.

When two or more pulses with displacements in opposite directions interact, the effect is known as **destructive interference**. The resulting displacements negate each other. Once the pulses have passed by each other, they continue along their original path in their original shape, as if they had never met. An animation of two pulses interfering constructively and destructively is available at http://bit.ly/hyJ3lZ.

8.27 Q: The diagram below represents two pulses approaching each other from opposite directions in the same medium.

Which diagram best represents the medium after the pulses have passed through each other?

(1)

(2)

(3)

(4)

8.27 A: (2) the pulses continue as if they had never met.

8.28 Q: The diagram below represents shallow water waves of constant wavelength passing through two small openings, A and B, in a barrier.

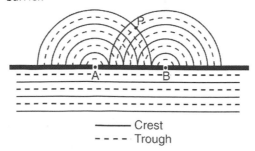

——— Crest
- - - - Trough

Which statement best describes the interference at point P?

(A) It is constructive, and causes a longer wavelength.

(B) It is constructive, and causes an increase in amplitude.

(C) It is destructive, and causes a shorter wavelength.

(D) It is destructive, and causes a decrease in amplitude.

8.28 A: (D) when a crest and a trough meet, destructive interference causes a decrease in amplitude.

8.29 Q: The diagram below shows two pulses of equal amplitude, A, approaching point P along a uniform string.

When the two pulses meet at P, the vertical displacement of the string at P will be

(A) A

(B) 2A

(C) 0

(D) A/2

8.29 A: (C) the pulses will experience destructive interference.

8.30 Q: The diagram below represents two pulses approaching each other.

Which diagram best represents the resultant pulse at the instant the pulses are passing through each other?

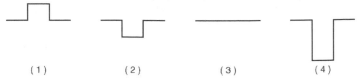

(1)   (2)   (3)   (4)

8.30 A: (2) shows the superposition (addition) of the two pulses.

8.31 Q: Two waves having the same amplitude and frequency are traveling in the same medium. Maximum destructive interference will occur when the phase difference between the waves is

(A) 0°
(B) 90°
(C) 180°
(D) 270°

8.31 A: (C) Maximum destructive interference occurs at a phase difference of 180°.

Standing Waves

When waves of the same frequency and amplitude traveling in opposite directions meet, a standing wave is produced. A **standing wave** is a wave in which certain points (**nodes**) appear to be standing still and other points (**antinodes**) vibrate with maximum amplitude above and below the axis. You can view an animation of this at http://aplusphysics.com/l/standwave.

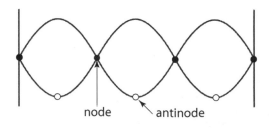

node   antinode

Looking at the standing wave produced above, you can see a total of four nodes in the wave, and three antinodes. For any standing wave pattern, you will always have one more node than antinode.

Standing waves can be observed in a variety of patterns and configurations, and are responsible for the functioning of most musical instruments. Guitar strings, for example, demonstrate a standing wave pattern. By fretting the strings, you adjust the wavelength of the string, and therefore, the wavelength of the standing waves that it can support, adjusting the frequency of the standing wave pattern, creating a different pitch. Similar functionality is seen in instruments ranging from pianos and drums to flutes, harps, trombones, xylophones, and even pipe organs!

8.32 Q: While playing, two children create a standing wave in a rope, as shown in the diagram below.

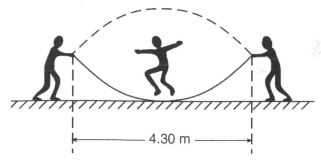

4.30 m

A third child participates by jumping the rope. What is the wavelength of this standing wave?

(A) 2.15 m

(B) 4.30 m

(C) 6.45 m

(D) 8.60 m

8.32 A: (D) the standing wave shown is half a wavelength, which implies the total wavelength must be 8.6m.

8.33 Q: Wave X travels eastward with frequency f and amplitude A. Wave Y, traveling in the same medium, interacts with wave X and produces a standing wave. Which statement about wave Y is correct?

(A) Wave Y must have a frequency of f, an amplitude of A, and be traveling eastward.

(B) Wave Y must have a frequency of 2f, an amplitude of 3A, and be traveling eastward.

(C) Wave Y must have a frequency of 3f, an amplitude of 2A, and be traveling westward.

(D) Wave Y must have a frequency of f, an amplitude of A, and be traveling westward.

8.33 A: (D) Standing waves are created when waves with the same frequency and amplitude traveling in opposite directions meet.

8.34 Q: The diagram below represents a wave moving toward the right side of this page.

Which wave shown below could produce a standing wave with the original wave?

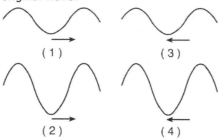

8.34 A: (3) must have same frequency, amplitude, and be traveling in the opposite direction in the same medium.

8.35 Q: Rank the following four standing waves in the same medium from highest to lowest in terms of A) frequency, B) period, C) wavelength, and D) amplitude.

8.35 A: A) Frequency: D, C, B, A
B) Period: A, B, C, D
C) Wavelength: A, B, C, D
D) Amplitude: C, B, D, A

8.36 Q: The diagram below shows a standing wave.

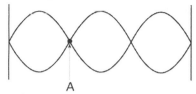

A

Point A on the standing wave is
(A) a node resulting from constructive interference
(B) a node resulting from destructive interference
(C) an antinode resulting from constructive interference
(D) an antinode resulting from destructive interference

8.36 A: (B) a node resulting from destructive interference.

8.37 Q: One end of a rope is attached to a variable speed drill and the other end is attached to a 5-kilogram mass. The rope is draped over a hook on a wall opposite the drill. When the drill rotates at a frequency of 20 Hz, standing waves of the same frequency are set up in the rope. The diagram below shows such a wave pattern.

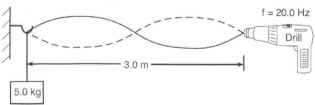

f = 20.0 Hz

Drill

3.0 m

5.0 kg

A) Determine the wavelength of the waves producing the standing wave pattern.

B) Calculate the speed of the wave in the rope.

8.37 A: A) Wavelength is 3.0 meters from diagram.

B) $v = f\lambda = (20\,Hz)(3m) = 60\,\frac{m}{s}$

Doppler Effect

The shift in a wave's observed frequency due to relative motion between the source of the wave and the observer is known as the **Doppler Effect**. In essence, when the source and/or observer are moving toward each other, the observer perceives a shift to a higher frequency, and when the source and/or observer are moving away from each other, the observer perceives a lower frequency.

This can be observed when a vehicle travels past you. As you hear the vehicle approach, you can observe a higher frequency noise, and as the vehicle passes by you and then moves away, you observe a lower frequency noise. This effect is the principle behind radar guns to measure an object's speed as well as meteorology radar which provides data on wind speeds.

The Doppler Effect results from waves having a fixed speed in a given medium. As waves are emitted, a moving source or observer encounters the wave fronts at a different frequency than the waves are emitted, resulting in a perceived shift in frequency. The video and animation at http://bit.ly/epLkPj may help you visualize this effect.

8.38 Q: A car's horn produces a sound wave of constant frequency. As the car speeds up going away from a stationary spectator, the sound wave detected by the spectator

(A) decreases in amplitude and decreases in frequency

(B) decreases in amplitude and increases in frequency

(C) increases in amplitude and decreases in frequency

(D) increases in amplitude and increases in frequency

8.38 A: (A) decreases in amplitude because the distance between source and observe is increasing, and decreases in frequency because the source is moving away from the observer.

8.39 Q: A car's horn is producing a sound wave having a constant frequency of 350 hertz. If the car moves toward a stationary observer at constant speed, the frequency of the car's horn detected by this observer may be

(A) 320 Hz

(B) 330 Hz

(C) 350 Hz

(D) 380 Hz

8.39 A: (D) If source is moving toward the stationary observer, the observed frequency must be higher than source frequency.

8.40 Q: A radar gun can determine the speed of a moving automobile by measuring the difference in frequency between emitted and reflected radar waves. This process illustrates

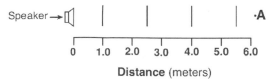

(A) resonance

(B) the Doppler effect

(C) diffraction

(D) refraction

8.40 A: (B) the Doppler effect.

8.41 Q: The vertical lines in the diagram represent compressions in a sound wave of constant frequency propagating to the right from a speaker toward an observer at point A.

Speaker →

| | | | ·A

0   1.0   2.0   3.0   4.0   5.0   6.0

Distance (meters)

A) Determine the wavelength of this sound wave.

B) The speaker is then moved at constant speed toward the observer at A. Compare the wavelength of the sound wave received by the observer while the speaker is moving to the wavelength observed when the speaker was at rest.

8.41 A: A) Wavelength is compression to compression, or 1.5m.

B) Observed frequency is higher while speaker is moving toward the observer due to the Doppler Effect, so the observed wavelength must be shorter.

8.42 Q: A student sees a train that is moving away from her and sounding its whistle at a constant frequency. Compared to the sound produced by the whistle, the sound observed by the student is

(A) greater in amplitude

(B) a transverse wave rather than a longitudinal wave

(C) higher in pitch

(D) lower in pitch

8.42 A: (D) lower in pitch since the source is moving away from the observer.

An exciting application of the Doppler Effect involves the analysis of radiation from distant stars and galaxies in the universe. Based on the basic elements that compose stars, scientists know what frequencies of radiation to look for. However, when analyzing these objects, they often observe frequencies shifted toward the red end of the electromagnetic spectrum (lower frequencies), known as the **Red Shift**. This indicates that these celestial objects must be moving away from the Earth. The more distant the object, the greater the red shift. Putting this together, you can conclude that more distant celestial objects are moving away from Earth faster than nearer objects, and therefore, the universe must be expanding!

8.43 Q: When observed from Earth, the wavelengths of light emitted by a star are shifted toward the red end of the electromagnetic spectrum. This Red Shift occurs because the star is

(A) at rest relative to Earth

(B) moving away from Earth

(C) moving toward Earth at decreasing speed

(D) moving toward Earth at increasing speed

8.43 A: (B) moving away from Earth.

Reflection

When a wave hits a boundary, three different events can occur. The wave may be:

- Reflected - wave bounces off a boundary
- Transmitted - wave is transmitted into the new medium
- Absorbed - energy of the wave is transferred into the boundary medium

The **law of reflection** states that the angle at which a wave strikes a reflective medium (the **angle of incidence**, or θ_i) is equal to the angle at which a wave reflects off the medium (the **angle of reflection**, or θ_r). Put more simply, $\theta_i = \theta_r$. In all cases, the angle of incidence and the angle of reflection are measured from a line perpendicular, or normal, to the reflecting surface.

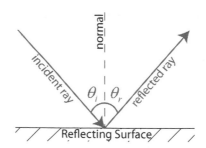

Although all waves can exhibit these behaviors, electromagnetic light waves are typically considered for demonstration purposes. When a wave bounces off a reflective surface, the nature of its reflection depends largely on the nature of the surface. Rough surfaces tend to reflect light in a variety of directions in a process known as diffuse reflection. **Diffuse reflection** is the type of reflection typically observed off of pieces of paper. Smooth surfaces tend to reflect light waves in a more regular fashion, such that the reflected rays maintain their parallelism. This process is known as **specular reflection**, and is commonly observed in mirrors.

8.44 Q: The diagram below represents a light ray striking the boundary between air and glass.

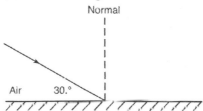

What would be the angle between this light ray and its reflected ray?

(A) 30°
(B) 60°
(C) 120°
(D) 150°

8.44 A: (C) recall that ray angles are measured to the normal, so the angle between the two rays is 60°+60°=120°.

8.45 Q: A sonar wave is reflected from the ocean floor. For which angles of incidence do the wave's angle of reflection equal its angle of incidence?

(A) angles less than 45°, only
(B) an angle of 45°, only
(C) angles greater than 45°, only
(D) all angles of incidence

(D) the law of reflection applies to all types of waves reflecting off a surface.

Mirrors

When you look in a flat (plane) mirror, you see a reflection, or an image, of an object. Light rays from the object reach the plane mirror and are reflected back to the observer, creating an image of the object. The image is known as a **virtual image** because the reflected rays don't actually pass through the image. All virtual images are upright.

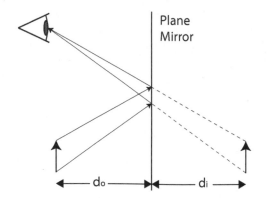

The distance from the object to the mirror is known as the **object distance** (d_o), and the distance from the image to the mirror is known as the **image distance** (d_i). Note that the AP exam typically refers to these as s_o and s_i. For virtual images, the image distance is negative. The **magnification** of an image is found using the magnification equation, which relates the image and object distances to the image (h_i) and object (h_o) heights.

$$m = \frac{-d_i}{d_o} = \frac{h_i}{h_o}$$

In the case of a plane mirror, the magnitude of the image distance is equal to the magnitude of the object distance. Therefore, the image appears the same size as the object.

8.46 Q: A student stands 2 meters in front of a vertical plane mirror. As the student walks toward the mirror, the image

(A) decreases in magnification and remains virtual

(B) decreases in magnification and remains real

(C) remains the same magnification and remains virtual

(D) remains the same magnification and remains real

8.46 A: (C) remains the same magnification and remains virtual since the image stays behind the reflecting surface, and the magnification of a plane mirror is always 1.

8.47 Q: Which diagram best represents image *I*, which is formed by placing object *O* in front of a plane mirror?

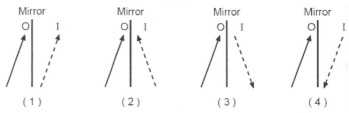

8.47 A: (2)

8.48 Q: In the diagram below, a person is standing 5 meters from a plane mirror. The chair in front of the person is located 2 meters from the mirror.

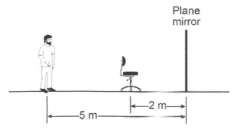

What is the distance between the person and the image he observes of the chair?

8.48 A: 7m

Not all mirrors are plane mirrors, however. The inner surface of a spherical **concave mirror** is reflective. Light rays coming into a mirror parallel to the **principal axis** (a virtual line perpendicular to the center of the mirror's surface) are reflected from the plane of the mirror and converge through the **focal point** of the mirror. Concave mirrors are also known as converging mirrors. The focal point of a spherical mirror is half its radius of curvature.

Light rays passing through the center of curvature strike the mirror and are reflected back through the center of curvature. Light rays from the object passing directly through the focal point are reflected back parallel to the principal axis. The convergence of the reflected rays creates an image. The distance from the focal point to the mirror's surface is known as the focal distance (f). This image is known as a **real image** because the reflected rays pass through the image. Real images are inverted.

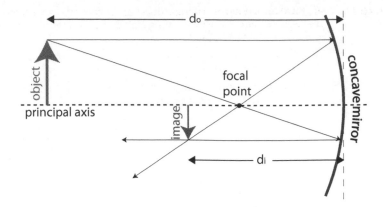

The relationship between the focal distance, the object distance, and the image distance is described by the mirror equation, also known as the lens equation. By convention, object and image distances are positive on the reflecting side of the mirror, and negative on the non-reflecting side of the mirror.

$$\frac{1}{f} = \frac{1}{d_o} + \frac{1}{d_i}$$

Analyzing an object inside the focal point of a concave mirror requires the same basic procedure. In this case, however, the reflected rays diverge on the reflective side of the mirror. To find the image, you must extend the real reflected rays back through the mirror onto the non-reflective side. The convergence of these extended rays leads to an upright, virtual image.

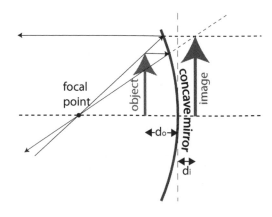

8.49 Q: An incident light ray travels parallel to the principal axis of a concave spherical mirror. After reflecting from the mirror, the light ray will travel

(A) through the mirror's focal point

(B) through the mirror's center of curvature

(C) parallel to the mirror's principal axis

(D) normal to the mirror's principal axis

8.49 A: (A) light rays parallel to the principal axis are reflected through the mirror's focal point.

8.50 Q: The diagram below shows an object located at point P, 0.25 meters from a concave spherical mirror with focal point F. The focal length of the mirror is 0.10 meters.

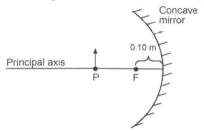

How does the image change as the object is moved from point P toward point F?

(A) Its distance from the mirror decreases and the size of the image decreases.

(B) Its distance from the mirror decreases and the size of the image increases.

(C) Its distance from the mirror increases and the size of the image decreases.

(D) Its distance from the mirror increases and the size of the image increases.

8.50 A: (D) Using the mirror equation, you observe that by moving the object toward point F, the object distance d_o decreases.

$$\frac{1}{f} = \frac{1}{d_o} + \frac{1}{d_i}$$

Since the focal point f is fixed, the image distance d_i must increase, so the image's distance from the mirror increases. Further, using the magnification equation, you observe that d_i increasing and d_o decreasing leads to an increase in magnification m.

$$m = \frac{-d_i}{d_o} = \frac{h_i}{h_o}$$

8.51 Q: An object arrow is placed in front of a concave mirror having center of curvature C and focal point F. Which diagram best shows the location of point I, the image of the tip of the object arrow? Is the image real or virtual? Upright or inverted?

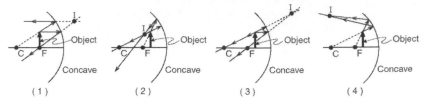

(1)   (2)   (3)   (4)

8.51 A: (3) The image is virtual and upright.

8.52 Q: A candle is placed 0.24 meters in front of a converging mirror that has a focal length of 0.12 meters. How far from the mirror is the image of the candle located?

(A) 0.08 m   (B) 0.12 m
(C) 0.24 m   (D) 0.36 m

8.52 A: (C) $\dfrac{1}{f} = \dfrac{1}{d_o} + \dfrac{1}{d_i} \rightarrow \dfrac{1}{d_i} = \dfrac{1}{f} - \dfrac{1}{d_o} \rightarrow$

$$\dfrac{1}{d_i} = \dfrac{1}{.12m} - \dfrac{1}{.24m} \rightarrow \dfrac{1}{d_i} = 4.17m^{-1} \rightarrow$$

$$d_i = \dfrac{1}{4.17m^{-1}} = 0.24m$$

The outer surface of a spherical **convex mirror** is reflective. Light rays coming into a convex mirror parallel to the principal axis are reflected away from the principal axis on a virtual line connecting the point of contact with the mirror plane and the focal point on the **non-reflecting** side of the mirror. For this reason, convex mirrors are also known as diverging mirrors. Light rays which strike the center of the mirror are reflected at the same angle. Because the light rays never converge on the reflective side of a convex mirror, convex mirrors only produce virtual images that are upright and reduced in size. Note that convex mirrors have negative focal distances.

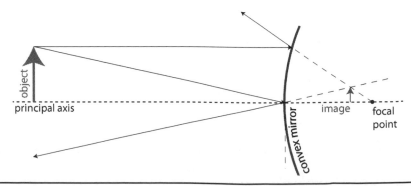

8.53 Q: Which optical device causes parallel light rays to diverge?
(A) convex mirror   (B) plane mirror
(C) concave mirror   (D) convex lens

8.53 A: (A) convex mirrors are also known as diverging mirrors.

8.54 Q: Images formed by diverging (convex) mirrors are always
(A) real and inverted
(B) real and erect
(C) virtual and inverted
(D) virtual and erect

8.54 A: (D) Images formed by diverging mirrors are virtual, erect, and reduced in size.

8.55 Q: Light rays from a candle flame are incident on a convex mirror. After reflecting from the mirror, these light rays
(A) converge and form a virtual image
(B) converge and form a real image
(C) diverge and form a virtual image
(D) diverge and form a real image

8.55 A: (C) Light rays incident on convex mirrors diverge and form only virtual images.

8.56 Q: The radius of curvature of a spherical mirror is R. The focal length of this mirror is equal to
(A) R/2   (B) 2R
(C) R/4   (D) 4R

8.56 A: (A) The focal length of a mirror is equal to half its radius of curvature.

Refraction

When a wave reaches the boundary between media, part of the wave is reflected and part of the wave enters the new medium. As the wave enters the new medium, the speed of the wave changes, and the frequency of a wave remains constant. Therefore, consistent with the wave equation, $v=f\lambda$, the wavelength of the wave must change.

> **8.57 Q:** When a wave enters a new material, what happens to its speed, frequency, and wavelength?

> **8.57 A:** Speed changes, frequency remains constant, and wavelength changes.

The front of a wave has some actual width, and if the wave does not impinge upon the boundary between media at a right angle, not all of the wave enters the new medium and changes speed at the same time. This causes the wave to bend as it enters a new medium in a process known as **refraction**.

To better illustrate this, imagine you're in a line in a marching band, connected with your bandmates as you march at a constant speed down the field in imitation of a wave front. As your wavefront reaches a new medium that slows you down, such as a mud pit, the band members reaching the mud pit slow down before those who reach the pit later. Since you are all connected in a wave front, the entire wave shifts directions (refracts) as it passes through the boundary between field and mud!

The **index of refraction** (n) is a measure of how much light slows down in a material. In a vacuum and in air, all electromagnetic waves have a speed of $c=3\times10^8$ m/s. In other materials, light slows down. The ratio of the speed of light in a vacuum to the speed of light in the new material is known as the index of refraction (n). The slower the wave moves in the material, the larger the index of refraction:

$$n = \frac{c}{v}$$

Not only does index of refraction depend upon the medium the light wave is traveling through, it also varies with frequency. This is known as **dispersion**. This frequency dependence is typically relative small. Dispersion is responsible for the behavior of prisms. White light, consisting of all visible frequencies, is refracted twice in a prism, first upon entering, and again upon leaving. The index of refraction of the prism material varies slightly with respect to frequency, causing the different frequencies of light to bend slightly different amounts.

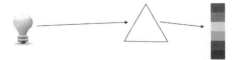

Absolute Indices of Refraction	
($f = 5.09 \times 10^{14}$ Hz)	
Air	1.00
Corn oil	1.47
Diamond	2.42
Ethyl alcohol	1.36
Glass, crown	1.52
Glass, flint	1.66
Glycerol	1.47
Lucite	1.50
Quartz, fused	1.46
Sodium chloride	1.54
Water	1.33
Zircon	1.92

8.58 Q: A light ray traveling in air enters a second medium and its speed slows to 1.71×10^8 m/s. What is the absolute index of refraction of the second medium?

8.58 A: $n = \dfrac{c}{v} = \dfrac{3 \times 10^8 \text{ m/}_s}{1.71 \times 10^8 \text{ m/}_s} = 1.75$

8.59 Q: In which way does blue light change as it travels from diamond into crown glass?

(A) Its frequency decreases.

(B) Its frequency increases.

(C) Its speed decreases.

(D) Its speed increases.

8.59 A: (D) Its speed increases because it crosses from a higher index of refraction material to a lower index of refraction material.

8.60 Q: Which characteristic is the same for every color of light in a vacuum?

(A) energy

(B) frequency

(C) speed

(D) period

8.60 A: (C) the speed of all EM waves in a vacuum is 3.0×10^8 m/s.

8.61 Q: A periodic wave travels at speed v through medium A. The wave passes with all its energy into medium B. The speed of the wave through medium B is v/2. Draw the wave as it travels through medium B.

8.61 A:
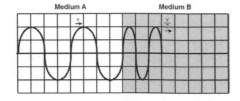

8.62 Q: A beam of monochromatic light has a wavelength of 5.89×10^{-7} meters in air. Calculate the wavelength of this light in diamond.

8.62 A:
$$\frac{n_2}{n_1}=\frac{\lambda_1}{\lambda_2}$$

$$\lambda_2=\frac{n_1\lambda_1}{n_2}=\frac{(1.00)(5.89\times10^{-7}\,m)}{2.42}=2.43\times10^{-7}\,m$$

8.63 Q: The speed of light in a piece of plastic is 2.00×10^8 meters per second. What is the absolute index of refraction of this plastic?
   (A) 1.00
   (B) 0.670
   (C) 1.33
   (D) 1.50

8.63 A: (D) $n=\dfrac{c}{v}=\dfrac{3\times10^8\,m/_s}{2\times10^8\,m/_s}=1.5$

Chapter 8: Optics

The amount a light wave bends as it enters a new medium is given by the law of refraction, also known as **Snell's Law**, which states:

$$n_1 \sin \theta_1 = n_2 \sin \theta_2$$

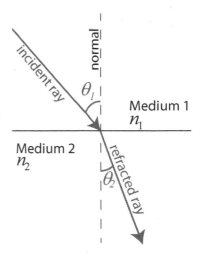

In this formula, n_1 and n_2 are the indices of refraction of the two media, and θ_1 and θ_2 correspond to the angles of the incident and refracted rays, again measured from the normal.

Light bends toward the normal as it enters a material with a higher index of refraction (slower material), and bends away from the normal as it enters a material with a lower index of refraction (faster material).

8.64 Q: A ray of light ($f=5.09\times10^{14}$ Hz) traveling in air is incident at an angle of 40° on an air-crown glass interface as shown below.

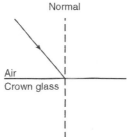

What is the angle of refraction for this light ray?

(A) 25°

(B) 37°

(C) 40°

(D) 78°

8.64 A: (A) $n_1 \sin \theta_1 = n_2 \sin \theta_2$

$$\theta_2 = \sin^{-1} \frac{n_1 \sin \theta_1}{n_2} = \sin^{-1} \frac{1.00 \times \sin 40°}{1.52} = 25°$$

8.65 Q: A ray of monochromatic light (f=5.09×10¹⁴ Hz) passes from air into Lucite at an angle of incidence of 30°.

A) Calculate the angle of refraction in the Lucite.

B) Using a protractor and straightedge, draw the refracted ray in the Lucite.

8.65 A: A) $n_1 \sin\theta_1 = n_2 \sin\theta_2$

$$\theta_2 = \sin^{-1}\frac{n_1 \sin\theta_1}{n_2} = \sin^{-1}\frac{1.00 \times \sin 30°}{1.50} = 19°$$

B)

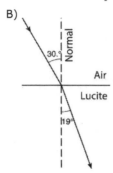

8.66 Q: When a light wave enters a new medium and is refracted, there must be a change in the light wave's

(A) color

(B) frequency

(C) period

(D) speed

8.66 A: (D) the change in a wave's speed causes its refraction.

8.67 Q: A ray of light (f=5.09×10¹⁴ Hz) traveling in air strikes a block of sodium chloride at an angle of incidence of 30°. What is the angle of refraction for the light ray in the sodium chloride?

(A) 19°

(B) 25°

(C) 40°

(D) 49°

8.67 A: (A) $n_1 \sin\theta_1 = n_2 \sin\theta_2$

$$\theta_2 = \sin^{-1}\frac{n_1 \sin\theta_1}{n_2} = \sin^{-1}\frac{1.00 \times \sin 30°}{1.54} = 19°$$

8.68 Q: Which diagram best represents the behavior of a ray of monochromatic light in air incident on a block of crown glass?

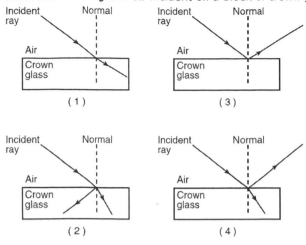

(1) (3)

(2) (4)

8.68 A: (4) shows both reflection and refraction of the incoming light ray.

When light passes from a higher-index (slower) material to a lower-index (faster) material, the light bends away from the normal. When the angle of refraction reaches 90°, the refracted ray would travel on the boundary between the surfaces. This occurs at an angle of incidence known as the **critical angle**, θ_c. For all angles of incidence greater than the critical angle, the light is reflected instead of refracted, known as **total internal reflection**. You can determine the critical angle from a brief derivation beginning with Snell's Law:

$$n_1 \sin\theta_1 = n_2 \sin\theta_2 \xrightarrow[\theta_2=90°]{\theta_1=\theta_c} n_1 \sin\theta_c = n_2 \rightarrow \theta_c = \sin^{-1}\left(\frac{n_2}{n_1}\right)$$

Total internal reflection is used in fiber topics to send optical signals tremendous distances with minimal intensity loss, and is also used in optical instruments such as telescopes, binoculars, fingerprint scanning, medical imaging devices. The brilliance of well-cut diamonds is also due to total internal reflection, where a vast majority of light incident upon the diamond is reflected back.

8.69 Q: Determine the critical angle for a light ray exiting from water (n=1.33) into air.

8.69 A: $\theta_c = \sin^{-1}\left(\frac{n_2}{n_1}\right) = \sin^{-1}\left(\frac{1}{1.33}\right) = 49°$

Thin Lenses

A common application of refraction is the optical lens. Much like mirrors, lenses come in two types: convex and concave. When working with lenses, however, convex lenses are converging lenses, and concave lenses are diverging lenses.

In the case of lenses, similar rules for ray tracing apply, and you can use the same equations. Convention states that image distances beyond the lens are considered positive distance, and images on the same side of the lens as the object are considered negative distance. For a convex lens, a ray parallel to the principal axis is refracted through the far focal point of the lens. In addition, a ray drawn from the object through the center of the lens passes through the center of the lens unbent (it actually refracts upon entering the lens and refracts back upon leaving, but the net result is often drawn as a straight line directly through the lens.)

Any lens that is thicker on the edges than the center acts as a diverging lens and any lens that is thicker in the center than the edges acts as a converging lens.

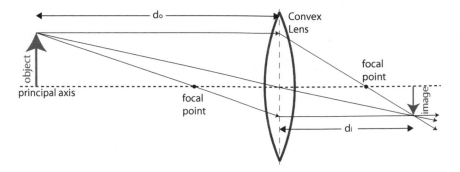

8.70 Q: The diagram below shows an arrow placed in front of a converging lens.

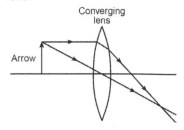

The lens forms an image of the arrow that is

(A) real and inverted

(B) real and erect

(C) virtual and inverted

(D) virtual and erect

8.70 A: (A) the image will be real and inverted.

8.71 Q: An object is located 0.15 meters from a converging lens with focal length 0.10 meters. How far from the lens is the image formed?

8.71 A:
$$\frac{1}{f} = \frac{1}{d_o} + \frac{1}{d_i} \rightarrow \frac{1}{d_i} = \frac{1}{f} - \frac{1}{d_o} \rightarrow$$

$$\frac{1}{d_i} = \frac{1}{.10m} - \frac{1}{.15m} \rightarrow \frac{1}{d_i} = 3.33m^{-1} \rightarrow$$

$$d_i = \frac{1}{3.33m^{-1}} = 0.3m$$

8.72 Q: A converging lens forms a real image that is four times larger than the object. If the image is located 0.16 meters from the lens, what is the object distance?

8.72 A: First, realize that since the image is real, the image distance, by convention, is negative. Therefore, you are given the magnification of m=4, and the image distance d_i=-0.16m. Now you can apply the magnification equation to solve for the object distance d_o.

$$m = \frac{-d_i}{d_o} \rightarrow d_o = \frac{-d_i}{m} = \frac{-(-0.16m)}{4} = 0.04m$$

For a concave lens, a ray from the object parallel to the principal axis is refracted away from the principal axis on a line from the near focal point through the point where the ray intercepts the center of the lens. In addition, any ray that passes from the object through a focal point is refracted parallel to the principal axis. This leads to upright, virtual, reduced images from concave diverging lenses.

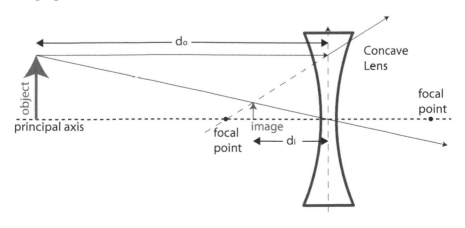

The same equations used for analysis of mirrors also apply to the analysis of thin lenses. Recall the lens equation:

$$\frac{1}{f} = \frac{1}{d_o} + \frac{1}{d_i}$$

Calculation of the magnification of a lens also uses the same equation used in the analysis of mirrors:

$$m = \frac{-d_i}{d_o} = \frac{h_i}{h_o}$$

8.73 Q: In the diagram below, parallel light rays in air diverge as a result of interacting with an optical device.

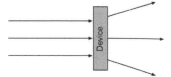

The device could be a

(A) convex glass lens

(B) rectangular glass block

(C) plane mirror

(D) concave glass lens

8.73 A: (D) a concave lens is also known as a diverging lens.

8.74 Q: Which glass lens in air can produce an enlarged real image of an object?

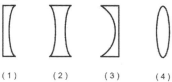

(1)   (2)   (3)   (4)

8.74 A: (4) convex lenses can produce enlarged real images.

8.75 Q: Which ray best represents the path of light ray R after it passes through the lens?

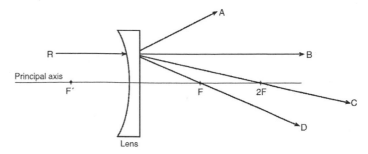

8.75 A: (A) shows the diverging light ray from a diverging (concave) lens.

8.76Q: The diagram shows an object placed between 1 and 2 focal lengths from a converging lens. The image of the object produced by the lens is

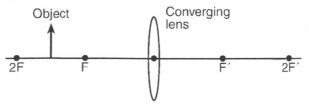

(A) real and inverted   (B) virtual and inverted

(C) real and erect   (D) virtual and erect

8.76 A: (A) real and inverted

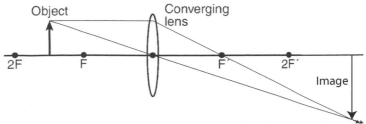

8.77 Q: A converging lens has a focal length of 0.08 meters. A light ray travels from the object to the lens parallel to the principal axis.

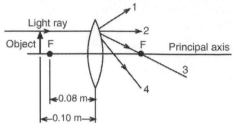

A) Which line best represents the path of the ray after it leaves the lens?

B) How far from the lens is the image formed?

C) Which phenomenon best explains the path of the light ray through the lens?
(1) diffraction   (2) dispersion
(3) reflection   (4) refraction

8.77 A: A) (3) Rays parallel to the principal axis are refracted through the far focal point of a convex lens.

B) $\dfrac{1}{f} = \dfrac{1}{d_o} + \dfrac{1}{d_i} \rightarrow \dfrac{1}{d_i} = \dfrac{1}{f} - \dfrac{1}{d_o} \rightarrow$

$\dfrac{1}{d_i} = \dfrac{1}{0.08m} - \dfrac{1}{0.10m} \rightarrow \dfrac{1}{d_i} = 2.5m^{-1} \rightarrow$

$d_i = \dfrac{1}{2.5m^{-1}} = 0.4m$

C) (4) refraction is responsible for the bending of light in lenses.

Diffraction

Diffraction is the bending of waves around obstacles, or the spreading of waves as they pass through an opening, most apparent when looking at obstacles or openings having a size of the same order of magnitude as the wavelength.

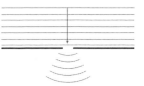

Typically, the smaller the obstacle and longer the wavelength, the greater the diffraction. Taken to the extreme, when a wave is blocked by a small enough opening, the wave passing through the opening actually behaves like a point source for a new wave.

8.78 Q: Which diagram best represents the shape and direction of a series of wave fronts after they have passed through a small opening in a barrier?

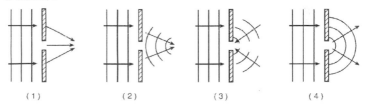

(1)   (2)   (3)   (4)

8.78 A: (4) the wave spreads out as it passes through a small opening.

8.79 Q: A beam of monochromatic light approaches a barrier having four openings, A, B, C, and D, of different sizes as shown below.

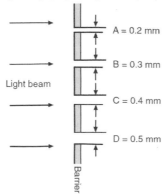

Which opening will cause the greatest diffraction?

8.79 A: (A) has the smallest opening, so will create the most diffraction.

8.80 Q: Parallel wave fronts incident on an opening in a barrier are diffracted. For which combination of wavelength and size of opening will diffraction effects be greatest?

(A) short wavelength and narrow opening

(B) short wavelength and wide opening

(C) long wavelength and narrow opening

(D) long wavelength and wide opening

8.80 A: (C) long wavelength and narrow opening produces the greatest diffraction. Wavelengths that are much shorter than the width of the opening produce only minimal diffraction effects.

8.81 Q: A wave of constant wavelength diffracts as it passes through an opening in a barrier. As the size of the opening is increased, the diffraction effects

(A) decrease

(B) increase

(C) remain the same

8.81 A: (A) As the size of the opening increases, the amount of diffraction decreases.

8.82 Q: The diagram below shows a plane wave passing through a small opening in a barrier.

Sketch four wave fronts after they have passed through the barrier.

8.82 A:

In 1678, Dutch physicist Christiaan Huygens described a method for analyzing the movement of wave fronts that helps explain diffraction. **Huygens' Principle** states that every point on a wave front acts as a source of spherical waves, which propagate at the speed of the wave itself. The sum, or superposition, of all the spherical wavefronts from all the individual points determines the new wave front. An illustration of Huygens' Principle applied to a wave incident upon a small opening is shown below.

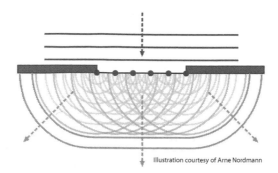

Illustration courtesy of Arne Nordmann

You can observe diffraction quite easily by analyzing monochromatic light passing through a narrow slit of width d before reaching a viewing screen arranged a length L away. As the light waves pass through the narrow slit, parallel rays pass directly through the center of the slit and reach the center of the screen, creating a central bright spot on the screen. Rays that are diffracted as they pass through the opening, however, travel a different distance to the screen depending on whether they were diffracted from the top of the opening or the bottom of the opening. This results in the waves reaching the viewing screen out of phase by varying amounts at different distances from the central bright spot. When they reach the viewing screen completely out of phase, you obtain a minimum of zero intensity in the pattern, known as a minima. This pattern repeats across the viewing screen, resulting in what is known as a single-slit diffraction pattern, illustrated below.

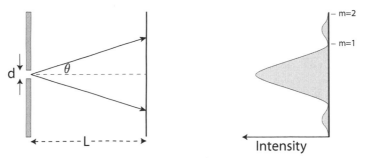

You can relate the slit width, the angle at which destructive interference occurs, the wavelength of the light, and the minima number (shown in the diagram) using the following diffraction equation:

$$d \sin \theta = m\lambda$$

The maxima, then, occur at m=0 (the central maximum), m=1.5, m=2.5, and so on.

Thomas Young's Double-Slit Experiment is a famous experiment which utilized diffraction to prove light has properties of waves. Young placed a light source behind a barrier with two narrow slits, allowing only a small portion of the light to pass through each slit.

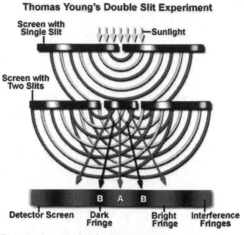

Illustration Courtesy of Michael W. Davidson

Because the two light waves travel different distances to the screen on which they are projected, you can see effects of both constructive and destructive interference, phenomena that occur only for waves. The interference pattern generated is similar to that of single-slit diffraction, though there are some important differences. First, the width of the central maximum is the same as the width of the other maxima, and the difference in intensity across the maxima is less extreme. Further, in the case of double-slit diffraction, the maxima occur at m=0, m=1, m=2, ..., whereas the minima occur at m=0.5, m=1.5, m=2.5, and so on when applying the diffraction equation dsinθ=mλ.

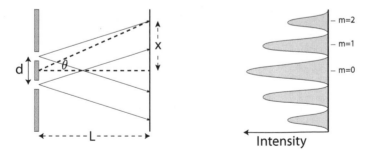

Going further, tan(θ) is equal to x/L, which is approximately equal to sin(θ) for small angles (and therefore approximately equal to x/L itself), so you can solve for the distance between bright spots as follows:

$$d \sin\theta = m\lambda \xrightarrow{\sin\theta=x/L} d\frac{x}{L} = m\lambda \to x = \frac{m\lambda L}{d}$$

When a large number of slits are arranged at equal spacings, a **diffraction grating** is created. Diffraction gratings create the same pattern as double-slit diffraction, but with sharper and brighter maxima. Diffraction gratings are useful tools for determining wavelengths accurately. In addition, when white light strikes a diffraction grating, a spectrum of the constituent colors is spread out across the viewing screen.

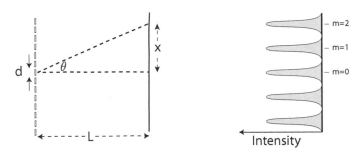

8.83 Q: Red light of wavelength 700 nm is incident upon a double slit of separation 0.0005 m.

A) What is the largest possible distance between a first-order and third-order bright spot viewed on a screen 1 meter from the slits?

B) What would happen to your answer if you replaced the double slit apparatus with a diffraction grating having the same slit separation?

8.83 A: A) The largest possible distance between a first-order and third-order bright spot will occur when the first-order bright spot is on the opposite side of the central maximum as the third-order bright spot, as shown below.

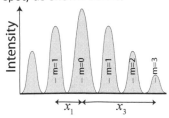

Next, find the distances x_1 and x_3 and add them to get the largest possible distance between a first-order and third-order bright spot.

$$x_1 = \frac{m\lambda L}{d} = \frac{1(700 \times 10^{-9} m)(1m)}{0.0005m} = 0.0014m$$

$$x_3 = \frac{m\lambda L}{d} = \frac{3(700 \times 10^{-9} m)(1m)}{0.0005m} = 0.0042m$$

$$x_{tot} = x_1 + x_3 = 0.0014m + 0.0042m = 0.0056m$$

B) Same spacing, but the spots would get sharper and brighter.

8.84 Q: Monochromatic radiation is incident upon a diffraction grating with 2000 lines/cm. If the distance between the central and first order bright spots on a screen 1 meter away is 11 cm, determine the wavelength of the radiation.

8.84 A: First determine the spacing between slits.

$$2000\,^{lines}/_{cm} \rightarrow \frac{0.01m}{2000lines} \rightarrow d = 5\times10^{-6}\,m$$

Next, solve for the wavelength of radiation.

$$x = \frac{m\lambda L}{d} \rightarrow \lambda = \frac{xd}{mL} = \frac{(0.11m)(5\times10^{-6}\,m)}{(1)(1m)} = 5.5\times10^{-7}\,m = 550nm$$

8.85 Q: Parallel rays of monochromatic light of given wavelength are incident upon an opaque film with various slit configurations as shown. A viewing screen is placed some distance away.

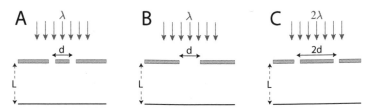

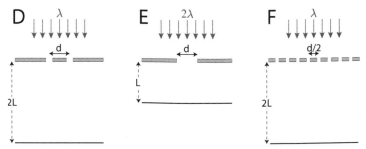

Rank the separation of the central peak to the next nearest maximum on the viewing screen from greatest to least.

8.85 A: F > E > D > B > A=C by application of the diffraction equation.

X-Ray Diffraction

Diffraction of x-ray radiation has provided a tremendously useful technique for exploring the composition of atomic-scale materials. A beam of x-rays incident upon a crystalline solid at some angle θ leads to reflection of the beam. Some of the beam will be reflected at the first plane of atoms, and some of the beam will be reflected at the next plane of atoms.

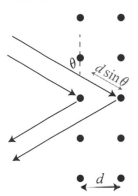

The two reflected rays interfere with each other, producing constructive maxima when the second ray travels a whole integer multiple of wavelengths greater than the first ray. The mathematical relationship describing the condition for maxima creation is:

$$2d \sin\theta = m\lambda$$

The maxima, then, occur at m=1, 2, 3, and so on. This is known as the Bragg Equation, and is used to better understand the atomic structure of atoms and molecules.

Thin Film Interference

When light traveling through a medium of index n_0 is Incident upon a thin film of thickness t and index of refraction n_1, some light is reflected (1st reflected ray) and some is transmitted into the film. If the transmitted light is again reflected at a boundary with another medium of index n_2, it travels back out of the film (2nd reflected ray) and can interfere with the first reflected ray as illustrated in the diagram below.

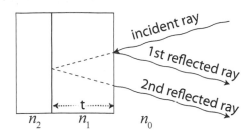

Depending on the materials and thicknesses involved, a variety of effects can be observed, which are used in applications such as anti-reflective coatings, dichroic color filters, ellipsometry (a method of determining the thickness and index of refraction of thin films), and other thin film metrology. You can also see the effects of thin film interference in everyday life, from the green tint of the anti-reflective coating on eyeglasses to the colors present in oil slicks and soap bubbles.

Analysis of thin film interference focuses on determining the optical path difference between the first reflected ray and the second reflected ray, noting that anytime reflection occurs off a slower medium (reflecting from a low-index material off a high-index material) a phase change occurs, resulting in an additional optical path difference of $\lambda_{film}/2$. Once the optical path difference between the two reflected rays has been determined, you find constructive interference with path differences of whole integer multiples of the wavelength in the film (0, λ_{film}, $2\lambda_{film}$, ...), and destructive interference halfway between the constructive wavelengths ($\lambda_{film}/2$, $3\lambda_{film}/2$, $5\lambda_{film}/2$, ...). Although it sounds complicated at first glance, actual problem solving is relatively straightforward with a little practice.

In Order, No Quarter

A memory device, "in order, no quarter," may be helpful in determining whether the incident light will reflect at an interface. If the three indices of refraction are in order, i.e. $n_1 < n_2 < n_3$, then light with wavelength (in the material) of 1/4 the thickness of the thin film will not reflect. In order, no quarter. Light with wavelength in the material of half the thickness of the thin film will reflect.

If the indices are out of order, then light with wavelength in the material of 1/4 the thickness of the thin film will reflect and light with wavelength of 1/2 the thickness of the thin film will not.

Special Thanks to Christopher Becke of Warhill High School for this tip!

8.86 Q: A quartz lens (n=1.46) is given a thin anti-reflective coating of tantalum pentoxide (n=2.15) to reduce normal reflections at 550 nm. Determine the minimum thickness of this coating to achieve optimal results.

8.86 A: Since the first reflected ray reflects once off a higher-index material, it will experience a phase shift resulting in an optical path difference of $\lambda_{film}/2$. The 2nd reflected ray travels an additional distance of 2t, and experiences no phase shift, resulting in an optical path difference of 2t (similar to the previous problem). The wavelength of the light ray in the tantalum pentoxide film can be found as:

$$\lambda_{film} = \frac{\lambda_{air}}{n_{film}} = \frac{550nm}{2.15} = 255.8nm$$

To minimize reflections, destructive interference is the goal, so set the optical path difference to $\lambda_{film}/2$ and solve for the thickness of the film.

$$OPD = 2t - \frac{\lambda_{film}}{2} = \frac{\lambda_{film}}{2} \rightarrow t = \frac{\lambda_{film}}{2} = \frac{255.8nm}{2} = 127.9nm$$

(Using the "In Order, No Quarter" approach, note that the n's are out of order. Since you want no reflection, you need a film with a thickness equal to half the light's wavelength in the film.)

8.87 Q: A 500-nm ray of light is normally incident upon a soap bubble of index of refraction 1.36.

A) Determine the minimum thickness of the soap bubble for constructive interference.

B) Determine the minimum thickness of the soap bubble for destructive interference.

8.87 A: First draw a diagram of the situation.

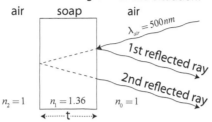

Since the first reflected ray reflects once off a higher-index material, it will experience a phase shift resulting in an optical path difference of $\lambda_{film}/2$. The 2nd reflected ray travels further than the 1st reflected ray by a distance of 2t, and experiences no phase shift, resulting in an optical path difference of 2t. You can also determine the wavelength of the light ray in the film as follows:

$$\lambda_{film} = \frac{\lambda_{air}}{n_{film}} = \frac{500nm}{1.36} = 368nm$$

This allows you to update your diagram with more specific information.

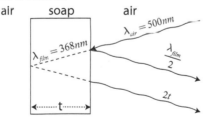

A) Constructive interference first occurs when the effective optical path difference between the two reflected rays is 0. Therefore you can solve for the thickness of the film at this condition:

$$OPD = 2t - \frac{\lambda_{film}}{2} = 0 \rightarrow t = \frac{\lambda_{film}}{4} = \frac{368nm}{4} = 92nm$$

B) Destructive interference first occurs when the effective optical path difference between the rays is $\lambda_{film}/2$:

$$OPD = 2t - \frac{\lambda_{film}}{2} = \frac{\lambda_{film}}{2} \rightarrow t = \frac{\lambda_{film}}{2} = \frac{368nm}{2} = 184nm$$

8.88 Q: A titanium dioxide bead (n=2.6) is given a thin anti-reflective coating of crown glass (n=1.52) to make it less reflective of 450-nm blue light at normal incidence. Determine the minimum thickness of the crown glass coating.

8.88 A: Since the first reflected ray reflects once off a higher-index material, it will experience a phase shift resulting in an optical path difference of $\lambda_{film}/2$. The 2nd reflected ray travels an additional distance of 2t, and also experiences a phase shift, resulting in an optical path difference of $2t + \lambda_{film}/2$. The wavelength of the light ray in the crown can be found as:

$$\lambda_{film} = \frac{\lambda_{air}}{n_{film}} = \frac{450nm}{1.52} = 296nm$$

To minimize reflections, the optical path difference between the reflected rays should be $\lambda_{film}/2$ or a whole integer multiple of that thickness. Solving for the thickness of the film:

$$OPD = \left(2t + \frac{\lambda_{film}}{2}\right) - \frac{\lambda_{film}}{2} = \frac{\lambda_{film}}{2} \rightarrow t = \frac{\lambda_{film}}{4} = \frac{296nm}{4} = 74nm$$

8.89 Q: Light is incident in air perpendicular to a thin film of glycerol (n=1.47) on top of water (n=1.33).

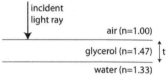

What minimum thickness of glycerol gives the reflected light a green (532 nm) color?

8.89 A: First find the wavelength of the green light in the glycerol film:

$$\lambda_{film} = \frac{\lambda_{air}}{n_{film}} = \frac{532nm}{1.47} = 362nm$$

Since the first reflected ray reflects once off a higher-index material, it will experience a phase shift resulting in an optical path difference of $\lambda_{film}/2$. The 2nd reflected ray travels an additional distance of 2t, but does not experience a phase shift, resulting in an optical path difference of 2t. Solve for the thickness of the film at constructive interference (optical path difference = 0):

$$OPD = 2t - \frac{\lambda_{film}}{2} = 0 \rightarrow t = \frac{\lambda_{film}}{4} = \frac{362nm}{4} = 90.5nm$$

Test Your Understanding

1. Explain how a liquid crystal display television works using your own words.

2. Explore why optical systems utilizing total internal reflection are typically more efficient than those using mirrors.

3. What is the focal length and magnification of a plane mirror?

4. Describe how fiber-optics can transmit large amounts of data quickly. Include references to total internal reflection.

5. Research the differences between optical systems for visible light and optical systems for x-rays. Why are these systems so different?

6. Design an experiment to determine the speed of light in a transparent block of an unknown material.

7. Explore how rainbows are formed. What conditions are required for you to see a rainbow?

8. Explain how x-ray diffraction can be used to measure atoms.

9. Make a table comparing the behaviors of a single slit, double slit, and diffraction grating when exposed to incoming parallel rays of monochromatic light.

10. Create a poster introducing the creation, function, and applications of holograms. Make sure to include references to interference patterns and diffraction.

Chapter 9: Modern Physics

"God does not play dice with the cosmos."
— *Albert Einstein*

"Einstein, don't tell God what to do."
— *Niels Bohr*

Objectives

1. Explain what is meant by wave-particle duality.
2. Provide evidence for both the wave and particle nature of light and matter.
3. Describe the Rutherford and Bohr models of the atom.
4. Calculate the energy of an absorbed or emitted photon from an energy level diagram.
5. Explain the quantum nature of atomic energy levels.
6. Compare and contrast emission, absorption, and continuous spectra.
7. Describe the quantum mechanical model of the atom.
8. Recognize that quantum mechanical processes are described by probability.
9. Recognize fundamental particles and forces consistent with the Standard Model of Particle Physics.
10. Describe various nuclear processes such as fission, fusion, and radioactive decay.
11. Understand and use the mass-energy equivalence equation.
12. Recognize situations in which classical physics must be replaced by special relativity to describe observations in various frames of reference.

Modern Physics refers largely to advancements in physics from the 1900s to the present, extending the models of Newtonian (classical) mechanics and electricity and magnetism to the extremes of the very small, the very large, the very slow and the very fast. Modern Physics can encompass a tremendous variety of topics, which will be explored briefly in this book. Key topics for this exploration include:

- the dual nature of electromagnetic radiation
- models of the atom
- sub-atomic structure
- radioactivity
- mass-energy equivalence
- relativity

Wave-Particle Duality

Although electromagnetic waves exhibit many characteristics and properties of waves, they can also exhibit some characteristics and properties of particles. These "particles" are called **photons**. Light (and all EM radiation), therefore, has a dual nature. At times, light acts like a wave, and at other times it acts like a particle.

Characteristics of light that indicate light behaves like a wave include:

- Diffraction
- Interference
- Doppler Effect
- Young's Double-Slit Experiment

Characteristics of light that indicate light also acts as a particle include:

- Blackbody Radiation
- Photoelectric Effect
- Compton Effect

9.01 Q: Light demonstrates the characteristics of
(A) particles, only
(B) waves, only
(C) both particles and waves
(D) neither particles nor waves

9.01 A: (C) both particles and waves.

9.02 Q: Which phenomenon provides evidence that light has a wave nature?

(A) emission of light from an energy-level transition in a hydrogen atom

(B) diffraction of light passing through a narrow opening

(C) absorption of light by a black sheet of paper

(D) reflection of light from a mirror

9.02 A: (B) diffraction is a phenomenon only applicable to waves.

Blackbody Radiation

The radiation emitted from a very hot object (known as **black-body radiation**) didn't align with physicists' understanding of light as a wave. Specifically, very hot objects emitted radiation in a specific spectrum of frequencies and intensities which varied with the temperature of the object. Hotter objects had higher intensities at lower wavelengths (toward the blue/UV end of the spectrum), and cooler objects emitted more intensity at higher wavelengths (toward the red/infrared end of the spectrum). Physicists expected that at very short wavelengths the energy radiated would become very large, in contrast to observed spectra. This problem was known as the ultraviolet catastrophe.

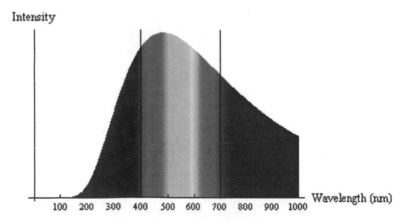

German physicist Max Planck solved this puzzle by proposing that atoms could only absorb or emit radiation in specific, non-continuous amounts, known as quanta. Energy, therefore, is quantized - it only exists in specific discrete amounts. For his work, Planck was awarded the Nobel Prize in Physics in 1918.

Photoelectric Effect

Further evidence that light behaves like a particle was proposed by Albert Einstein in 1905. Scientists had observed that when EM radiation struck a piece of metal, electrons could be emitted (known as **photoelectrons**). What was troubling was that not all EM radiation created photoelectrons. Regardless of what intensity of light was incident upon the metal, the only variable that affected the creation of photoelectrons was the frequency of the light.

If energy exists only in specific, discrete amounts, EM radiation exists in specific discrete amounts, and these smallest possible "pieces" of EM radiation are known as **photons**. A photon has zero mass and zero charge, and because it is a type of EM radiation, its velocity in a vacuum is equal to c (3×10^8 m/s). The energy of each photon of light is therefore quantized and is related to its frequency by the equation:

$$E_{photon} = hf = \frac{hc}{\lambda}$$

In this equation, the value of h, known as **Planck's Constant**, is given as 6.63×10^{-34} J•s.

Einstein proposed that the electrons in the metal object were held in an "energy well," and had to absorb at least enough energy to pull the electron out of the energy well in order to emit a photoelectron. The electrons in the metal would not be released unless they absorbed a single photon with that minimum amount of energy, known as the work function (ϕ) of the metal. The frequency of this photon is known as the **cutoff frequency**, or **threshold frequency**, f_0, of the metal. Any excess absorbed energy beyond that required to free the electron became kinetic energy for the photoelectron.

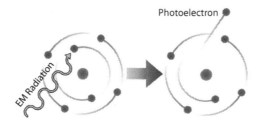

When a high-energy photon of light with energy greater than the energy holding an electron to its nucleus is absorbed by an atom, the electron is emitted as a photoelectron. The kinetic energy of the emitted photoelectron is exactly equal to the amount of energy holding the electron to the nucleus (the work function) subtracted from the energy of the absorbed photon.

$$K_{max} = hf - \phi$$

Chapter 9: Modern Physics

This theory extended Planck's work and inferred the particle-like behavior of photons of light. Photoelectrons would be ejected from the metal only if they absorbed a photon of light with frequency greater than or equal to a minimum threshold frequency, corresponding to the energy of a photon equal to the metal's "electron well" energy for the most loosely held electrons. Regardless of the intensity of the incident EM radiation, only EM radiation at or above the threshold frequency could produce photoelectrons.

In actually performing this experiment, the energy of the emitted photoelectrons is measured by collecting the electrons in a circuit. A reverse bias potential (negative potential difference) is then applied to the circuit until the current just becomes zero. This potential, known as the stopping potential, is the maximum kinetic energy of the emitted photoelectrons divided by the elementary charge.

$$V_{stopping} = \frac{K_{max}}{q}$$

In an experimental setting, creating a graph of the maximum kinetic energy of the emitted photoelectrons versus the frequency of the incident photons provides a tremendous amount of information, especially if you extend the linear function back to the y-axis in order to find the y-intercept.

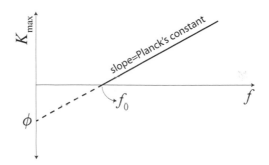

The slope of the line is equal to Planck's constant (h), the y-intercept gives you the work function of the metal (φ), and the x-intercept is the cut-off frequency (f_0).

9.03 Q: A photon of light traveling through space with a wavelength of 6×10^{-7} meters has an energy of

(A) 4.0×10^{-40} J
(B) 3.3×10^{-19} J
(C) 5.4×10^{10} J
(D) 5.0×10^{14} J

9.03 A: (B) $E = \dfrac{hc}{\lambda} = \dfrac{(6.63\times10^{-34}\,J\bullet s)(3\times10^{8}\,{}^{m}\!/\!_{s})}{6\times10^{-7}\,m} = 3.3\times10^{-19}\,J$

9.04 Q: The graph below represents the relationship between the energy and the frequency of photons.

Energy vs. Frequency

The slope of the graph would be

(A) 6.63×10^{-34} J•s

(B) 6.67×10^{-11} N•m²/kg²

(C) 1.60×10^{-19} J

(D) 1.60×10^{-19} C

9.04 A: (A) The slope of the graph, rise over run, is equivalent to the energy divided by the frequency, which gives you Planck's constant.

9.05 Q: The spectrum of visible light emitted during transitions in excited hydrogen atoms is composed of blue, green, red, and violet lines. What characteristic of light determines the amount of energy carried by a photon of that light?

(A) amplitude

(B) frequency

(C) phase

(D) velocity

9.05 A: (B) frequency determines the energy carried by a photon.

9.06 Q: Monochromatic light incident upon a photoelectric metal results in photoelectron emission. Which of the following could potentially halt the emission of photoelectrons? Select two answers.

(A) Increase the wavelength of incident light.

(B) Decrease the wavelength of incident light.

(C) Increase the frequency of incident light.

(D) Decrease the frequency of incident light.

9.06 A: (A) and (D). Both A and D reduce the energy of the incident photons, which could potentially reduce the energy of the incident photons below the work function of the metal and halt photoelectron emission.

Chapter 9: Modern Physics

9.07 Q: Determine the frequency of a photon whose energy is 3×10⁻¹⁹ joule.

9.07 A: $E = hf$

$$f = \frac{E}{h} = \frac{3 \times 10^{-19} J}{6.63 \times 10^{-34} J \bullet s} = 4.5 \times 10^{14} Hz$$

9.08 Q: Light of wavelength 5.0×10⁻⁷ meter consists of photons having an energy of
(A) 1.1×10⁻⁴⁸ J
(B) 1.3×10⁻²⁷ J
(C) 4.0×10⁻¹⁹ J
(D) 1.7×10⁻⁵ J

9.08 A: (C) $E = \dfrac{hc}{\lambda} = \dfrac{(6.63 \times 10^{-34} J \bullet s)(3 \times 10^{8}\,{}^{m}\!/_{s})}{5 \times 10^{-7} m} = 4 \times 10^{-19} J$

9.09 Q: The alpha line in the Balmer series of the hydrogen spectrum consists of light having a wavelength of 6.56×10⁻⁷ meter.
A) Calculate the frequency of this light.
B) Determine the energy in joules of a photon of this light.
C) Determine the energy in electronvolts of a photon of this light.

9.09 A: A) $v = f\lambda$

$$f = \frac{v}{\lambda} = \frac{3 \times 10^{8}\,{}^{m}\!/_{s}}{6.56 \times 10^{-7} m} = 4.57 \times 10^{14} Hz$$

B) $E = \dfrac{hc}{\lambda} = \dfrac{(6.63 \times 10^{-34} J \bullet s)(3 \times 10^{8}\,{}^{m}\!/_{s})}{6.56 \times 10^{-7} m} = 3.03 \times 10^{-19} J$

C) $3.03 \times 10^{-19} J \times \dfrac{1eV}{1.6 \times 10^{-19} J} = 1.89 eV$

9.10 Q: A photon has a wavelength of 9×10^{-10} meters. Calculate the energy of this photon in joules.

9.10 A: $E = \dfrac{hc}{\lambda} = \dfrac{(6.63 \times 10^{-34} \, J \bullet s)(3 \times 10^{8} \, ^{m}\!/_{s})}{9 \times 10^{-10} \, m} = 2.2 \times 10^{-16} \, J$

9.11 Q: Monochromatic light of a given frequency is incident upon various photoelectric materials, ejecting photoelectrons with a given velocity as shown below.

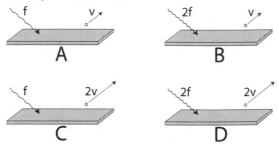

A) Which material has the greatest work function?

B) Which material has the smallest work function?

9.11 A: A) B has the greatest work function.

B) C has the smallest work function.

B has the greatest energy input due to the high frequency incoming photon, and the least energy out of the system due to the low-speed emitted photoelectron. Therefore, B must have the largest amount of energy required to liberate the photoelectron, and subsequently, the greatest work function. Using the same reasoning, C has the least energy input, coupled with the greatest energy output, therefore it must have the smallest work function.

9.12 Q: Monochromatic EM radiation of a specific wavelength is incident upon an unknown photoelectric material. The incident wavelength is varied and the stopping potential of the emitted photoelectrons is recorded as shown in the table below.

λ (nm)	f (Hz)	V_{stop} (V)	KE_{max} (eV)
265		2.85	
294		2.39	
352		1.70	
436		1.02	
520		0.56	

A) Complete the table.

B) Graph the maximum photoelectron kinetic energy vs. the frequency of the incident photons.

C) Determine the threshold frequency of the unknown material.

D) Determine the work function of the unknown material.

E) Use the graph to calculate Planck's constant. Compare this value to the accepted value and calculate your percent error.

9.12 A: A)

λ (nm)	f (Hz)	V_stop (V)	KE_max (eV)
265	1.13×10^{15}	2.85	2.85
294	1.02×10^{15}	2.39	2.39
352	8.52×10^{14}	1.70	1.70
436	6.88×10^{14}	1.02	1.02
520	5.77×10^{14}	0.56	0.56

B)

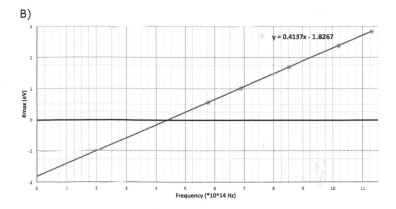

C) $f_0 = 4.4 \times 10^{14}$ Hz (x-intercept)

D) $\phi = 1.83$ eV (y-intercept)

E) Make sure you convert your units carefully!

$$h_{exp} = slope = \frac{0.4137 eV}{10^{14} Hz} \times \frac{1.6 \times 10^{-19} J}{1 eV} = 6.62 \times 10^{-34} J \cdot s$$

$$\% err = \frac{\left| h_{exp} - h_{acc} \right|}{h_{acc}} \times 100\% \rightarrow$$

$$\% err = \frac{\left| 6.62 \times 10^{-34} J \cdot s - 6.63 \times 10^{-34} J \cdot s \right|}{6.62 \times 10^{-34} J \cdot s} \times 100\% \rightarrow$$

$$\% err = 0.15\%$$

de Broglie Wavelength

Einstein continued to extend his theories around the interaction of photons and atomic particles, going so far as to hypothesize that photons could have momentum, also a particle property, even though they had no mass.

In 1922, American physicist Arthur Compton shot an X-ray photon at a graphite target to observe the collision between the photon and one of the graphite atom's electrons. Compton observed that when the X-ray photon collided with an electron, a photoelectron was emitted, but the original X-ray was also scattered and emitted, and with a longer wavelength (indicating it had lost energy).

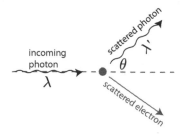

Further, the longer wavelength also indicated that the photon must have lost momentum. A detailed analysis showed that the energy and momentum lost by the X-ray was exactly equal to the energy and momentum gained by the photoelectron. Compton therefore concluded that not only do photons have momentum, they also obey the laws of conservation of energy and conservation of momentum! Compton won the Nobel Prize in Physics in 1927 for his work.

In 1923, French physicist Louis de Broglie took Compton's finding one step further. He stated that if EM waves can behave as moving particles, it would only make sense that a moving particle should exhibit wave properties. De Broglie's hypothesis was confirmed by shooting electrons through a double slit, similar to Young's Double-Slit Experiment, and observing a diffraction pattern. The smaller the particle, the more apparent its wave properties are. The wavelength of a moving particle, now known as the **de Broglie Wavelength**, is given by:

$$\lambda = \frac{h}{p}$$

Clinton Joseph Davisson and Lester Germer followed up on this work by shooting slow-moving electrons at a crystalline nickel target. When the reflected electrons were analyzed, they created the same diffraction pattern as observed when shooting x-rays at crystals (Bragg Diffraction). This experiment was verified by George Paget Thomson, and confirmed that matter behaves like waves. Davisson and Thomson shared the Nobel Prize in Physics in 1937 for this work.

                Chapter 9: Modern Physics

9.13 Q: Moving electrons are found to exhibit properties of
(A) particles, only
(B) waves, only
(C) both particles and waves
(D) neither particles nor waves

9.13 A: (C) moving particles have both particle and wave properties.

9.14 Q: Which phenomenon best supports the theory that matter has a wave nature?
(A) electron momentum
(B) electron diffraction
(C) photon momentum
(D) photon diffraction

9.14 A: (B) The diffraction of electrons indicates that electrons behave like waves.

9.15 Q: The kinetic energy of an electron is halved. What happens to the electron's de Broglie wavelength?
(A) The wavelength is decreased by a factor of $\sqrt{2}$.
(B) The wavelength is halved.
(C) The wavelength is increased by a factor of $\sqrt{2}$.
(D) The wavelength is doubled.

9.15 A: (C) The wavelength is increased by a factor of $\sqrt{2}$. Cutting the kinetic energy in half results in a new speed of $v/\sqrt{2}$. The de Broglie wavelength is inversely related to the speed, so the de Broglie wavelength is increased by a factor of $\sqrt{2}$.

9.16 Q: Wave-particle duality is most apparent in analyzing the motion of
(A) a baseball
(B) a space shuttle
(C) a galaxy
(D) an electron

9.16 A: (D) Wave-particle duality is most easily observed for small particles, especially evident when the de Broglie wavelength of the particle is smaller than the key features of the objects it interacts with.

9.17 Q: A particle with speed v is found to have a wavelength λ. What is the mass of a particle with twice the wavelength and twice the speed of the original?

(A) 4m

(B) 2m

(C) m/2

(D) m/4

9.17 A: (D) m/4. Mass is inversely related to both wavelength and speed; therefore, doubling the wavelength and doubling the speed requires a particle with one quarter the mass of the original.

Models of the Atom

In the early 1900s, scientists around the world began to refine and revise our understanding of atomic structure and sub-atomic particles. Scientists understood that matter was made up of atoms, and J.J. Thompson had shown that atoms contained very small negative particles known as electrons, but beyond that, the atom remained a mystery.

New Zealand scientist Ernest Rutherford devised an experiment to better understand the rest of the atom. The experiment, known as **Rutherford's Gold Foil Experiment**, involved shooting alpha particles (helium nuclei) at a very thin sheet of gold foil and observing the deflection of the particles after passing through the gold foil.

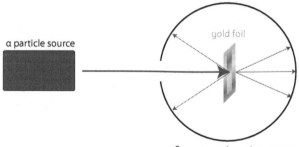

Rutherford and his graduate students, Geiger and Marsden, found that although most of the particles went through undeflected, a significant number of alpha particles were deflected by large amounts. Using an analysis based around Coulomb's Law and the conservation of momentum, Rutherford concluded that:

1. Atoms have a small, massive, positive nucleus at the center.
2. Electrons must orbit the nucleus.
3. Most of the atom is made up of empty space.

Rutherford's model was incomplete, though, in that it didn't account for a number of effects predicted by classical physics. Classical physics predicted that if the electron orbits the atom, it is constantly accelerating, and should therefore emit photons of EM radiation. Because the atom emits photons, it should be losing energy; therefore, the orbit of the electron would quickly decay into the nucleus and the atom would be unstable. Further, elements were found to emit and absorb EM radiation only at specific frequencies, which did not correlate to Rutherford's theory.

Following Rutherford's discovery, Danish physicist Niels Bohr traveled to England to join Rutherford's research group and refine Rutherford's model of the atom. Instead of focusing on all atoms, Bohr confined his research to developing a model of the simple hydrogen atom. Bohr's model made the following assumptions:

1. Electrons don't lose energy as they accelerate around the nucleus. Instead, energy is quantized. Electrons can only exist at specific discrete energy levels. These energy levels are given by the following equation, where Z is the atomic number and n is the positive integer energy level: $E_n = \dfrac{Z^2(-13.6eV)}{n^2}$

2. Each atom allows only a limited number of specific orbits (electrons) at each energy level.

3. To change energy levels, an electron must absorb or emit a photon of energy exactly equal to the difference between the electron's initial and final energy levels: $E_{photon} = E_i - E_f$.

Bohr's Model, therefore, was able to explain some of the limitations of Rutherford's Model. Further, Bohr was able to use his model to predict the frequencies of photons emitted and absorbed by hydrogen, explaining Rutherford's problem of emission and absorption spectra! For his work, Bohr was awarded the Nobel Prize in Physics in 1922.

9.18 Q: Calculate the energy and wavelength of the emitted photon when an electron moves from an energy level of -1.51 eV to -13.6 eV.

9.18 A: $E_{photon} = E_i - E_f = (-1.51eV) - (-13.6eV) = 12.09eV$

$$E_{photon} = \frac{hc}{\lambda}$$

$$\lambda = \frac{hc}{E_{photon}} = \frac{(6.63 \times 10^{-34} J \bullet s)(3 \times 10^8 \text{ }^m/_s)}{(12.09eV)(1.6 \times 10^{-19} \text{ }^J/_{eV})} = 1.03 \times 10^{-7} m$$

Energy Level Diagrams

A useful tool for visualizing the allowed energy levels in an atom is the energy level diagram. Two of these diagrams (one for hydrogen and one for mercury) are provided for you below. In each of these diagrams, the n=1 energy state is the lowest possible energy for an electron of that atom, known as the ground state. The energy corresponding to n=1 is shown on the right side of the diagram in electronvolts. So, for hydrogen, the ground state is a level of -13.6 eV.

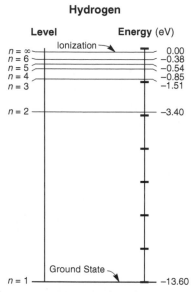

Energy Levels for the Hydrogen Atom

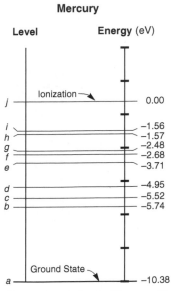

A Few Energy Levels for the Mercury Atom

The energy levels are negative to indicate that the electron is bound by the nucleus of the atom. If the electron reaches 0 eV, it is no longer bound by the atom and can be emitted as a photoelectron (i.e. the atom becomes ionized). Any remaining energy becomes the kinetic energy of the photoelectron.

9.19 Q: An electron in a hydrogen atom drops from the n=3 to the n=2 state. Determine the energy of the emitted radiation.

9.19 A:

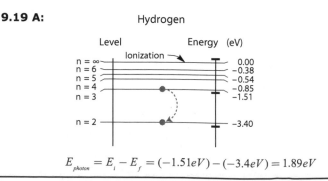

$$E_{photon} = E_i - E_f = (-1.51eV) - (-3.4eV) = 1.89eV$$

9.20 Q: Which type of photon is emitted when an electron in a hydrogen atom drops from the n = 2 to the n = 1 energy level?

(A) ultraviolet

(B) visible light

(C) infrared

(D) radio wave

9.20 A: (A) First find the amount of energy emitted in electron volts, convert that energy to Joules, then find the frequency of the emitted radiation, which you can look up on the EM Spectrum to determine the radiation type.

$$E_{photon} = E_i - E_f = (-3.4eV) - (-13.6eV) = 10.2eV$$

$$10.2eV \times \frac{1.6 \times 10^{-19} J}{1eV} = 1.63 \times 10^{-18} J$$

$$E = hf \rightarrow f = \frac{E}{h} = \frac{1.63 \times 10^{-18} J}{6.63 \times 10^{-34} J \bullet s} = 2.46 \times 10^{15} Hz$$

9.21 Q: Base your answers on the Energy Level Diagram for Hydrogen on the previous page.

A) Determine the energy, in electronvolts, of a photon emitted by an electron as it moves from the n = 6 to the n = 2 energy level in a hydrogen atom.

B) Convert the energy of the photon to joules.

C) Calculate the frequency of the emitted photon.

D) Is this the only energy and/or frequency that an electron in the n = 6 energy level of a hydrogen atom could emit? Explain your answer.

9.21 A: A) $E_{photon} = E_i - E_f = (-0.38eV) - (-3.4eV) = 3.02eV$

B) $3.02eV \times \frac{1.6 \times 10^{-19} J}{1eV} = 4.83 \times 10^{-19} J$

C) $E = hf \rightarrow f = \frac{E}{h} = \frac{4.83 \times 10^{-19} J}{6.63 \times 10^{-34} J \bullet s} = 7.29 \times 10^{14} Hz$

D) No, this is not the only energy and/or frequency that an electron in the n=6 energy level of a hydrogen atom could emit. The electron can return to any of the five lower energy levels.

9.22 Q: An electron in a mercury atom drops from energy level f to energy level c by emitting a photon having an energy of

(A) 8.20 eV

(B) 5.52 eV

(C) 2.84 eV

(D) 2.68 eV

9.22 A: (C) $E_{photon} = E_i - E_f = (-2.68eV) - (-5.52eV) = 2.84eV$

9.23 Q: A mercury atom in the ground state absorbs 20 electronvolts of energy and is ionized by losing an electron. How much kinetic energy does this electron have after the ionization?

(A) 6.40 eV

(B) 9.62 eV

(C) 10.38 eV

(D) 13.60 eV

9.23 A: (B) The ionization energy for an electron in the ground state of a mercury atom is 10.38 eV according to the Mercury Energy Level Diagram. If the atom absorbs 20 eV of energy, and uses up 10.38 eV in ionizing the electron, the electron has a leftover energy of 9.62 eV, which must be the electron's kinetic energy.

9.24 Q: A hydrogen atom with an electron initially in the n = 2 level is excited further until the electron is in the n = 4 level. This energy level change occurs because the atom has

(A) absorbed a 0.85-eV photon

(B) emitted a 0.85-eV photon

(C) absorbed a 2.55-eV photon

(D) emitted a 2.55-eV photon

9.24 A: (C) absorbed a 2.55-eV photon.

Atomic Spectra

Once you understand the energy level diagram, it quickly becomes obvious that atoms can only emit certain frequencies of photons, correlating to the difference between energy levels as an electron falls from a higher energy state to a lower energy state. In similar fashion, electrons can only absorb photons with energy equal to the difference in energy levels as the electron jumps from a lower to a higher energy state. This leads to unique atomic spectra of emitted and absorbed radiation for each element.

X-rays are produced using this concept. Electrons are accelerated through a large potential difference and collide with a molybdenum or platinum plate. The accelerated electrons collide with molybdenum or platinum electrons in the n=1 level. Electrons from higher energy levels then drop into the n=1 level, giving off X-ray photons.

An object that is heated to the point where it glows (**incandescence**) emits a continuous energy spectrum, known as blackbody radiation, described previously.

If a gas-discharge lamp is made from mercury vapor, the mercury vapor is made to emit light by application of a high electrical potential. The light emitted by the mercury vapor is created by electrons in higher energy states falling to lower energy states; therefore, the photons emitted correspond directly in wavelength to the difference in energy levels of the electrons. This creates a unique spectrum of frequencies which can be observed by separating the colors using a prism, known as an emission spectrum. By analyzing the emission spectra of various objects, scientists can determine the composition of those objects.

In similar fashion, if light of all colors is shone through a cold gas, the gas will only absorb the frequencies corresponding to photon energies exactly equal to the difference between the gas's atomic energy levels. This creates a spectrum with all colors except those absorbed by the gas, known as an absorption spectrum.

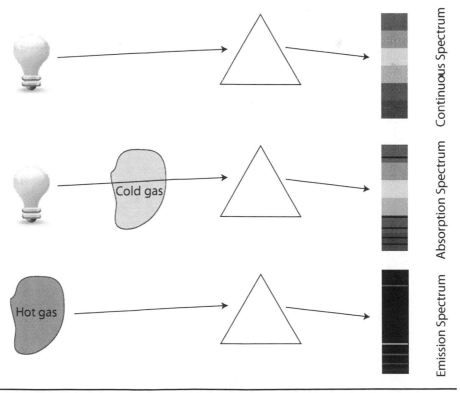

9.25 Q: The bright-line emission spectrum of an element can best be explained by

(A) electrons transitioning between discrete energy levels in the atoms of that element

(B) protons acting as both particles and waves

(C) electrons being located in the nucleus

(D) protons being dispersed uniformly throughout the atoms of that element

9.25 A: (A) bright-line emission spectra are created by electrons moving between energy levels, giving off photons of energy equal to the difference in energy levels.

9.26 Q: The diagram below represents the bright-line spectra of four elements, A, B, C, and D, and the spectrum of an unknown gaseous sample.

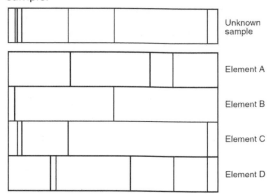

Based on comparisons of these spectra, which two elements are found in the unknown sample?

(A) A and B

(B) A and D

(C) B and C

(D) C and D

9.26 A: (C) Elements B and C have bright lines corresponding to the unknown sample.

Quantum Mechanical Model of the Atom

Bohr's model of the hydrogen atom was a tremendous step forward in understanding and describing the fundamental structure of matter, but it was not without flaws. For one thing, it didn't explain the relative brightness of spectral lines, nor the fact that a very close inspection of spectral lines showed that emission lines were actually comprised of a couple of very closely spaced lines. Further, it didn't predict the characteristics of more complex atoms well, and left many gaps in terms of how atoms bond to each other to create more complex structures. A number of advances beginning in the 1920s led to the development of the quantum mechanical model of the atom, which addressed a number of these shortcomings.

Like the Bohr model, the quantum mechanical model states that electrons exist in specific states, and that photons are emitted and absorbed in conjunction with electrons moving between states. However, electrons do not exist in orbits, but rather in electron clouds, which can be thought of like a probability distribution for a particle, or like a wave spread out in three-dimensional space. There are areas in the cloud where electrons are more likely to be observed, and areas where they are less likely to be observed, consistent with wave-particle duality.

Further, electron states are described by four characteristics, or quantum numbers. The first quantum number, the **principal quantum number** (n), relates to the energy of the electron state, and can be any whole number integer from one to infinity. The second quantum number, the **orbital quantum number** (l), is related to the angular momentum of the electron in the state, and can range from 0 to n-1 in whole number integers. The third quantum number, the **magnetic quantum number** (m_l), relates to the direction of the angular momentum. The final quantum number, the **spin quantum number** (m_s), is typically written as +½ or -½, though m_s is often called just spin up or spin down. The **Pauli Exclusion Principle** states that no two electrons can exist in exactly the same state.

If you think about electrons as standing waves about the nucleus, known as **de Broglie Waves**, the circumference of the orbital radius must be a whole integer multiple of the de Broglie wavelength of the particle, where higher energy levels (and principal quantum numbers) correspond to a greater circumference for the wave:

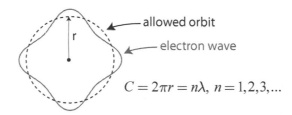

$$C = 2\pi r = n\lambda, \; n = 1, 2, 3, ...$$

This model of the atom assisted with understanding energy levels, but still left gaps in the understanding of more complex atoms.

Probabilistic Description of Matter

The de Broglie wavelength tells you the wavelength of a particle as it relates to the particle's momentum. Another important characteristic of waves, however, is their amplitude. The amplitude of a particle wave is given by the **wave function**, Ψ, which is a function of both position and time.

Actual calculation of the wave function in what is known as **Schrödinger's Equation** is well beyond the scope of this course, but you can begin to grasp its importance by looking at what it represents. Instead of looking at Ψ itself, you can interpret the value of Ψ^2 as the probability of finding a particle at the given position and time. If the value of Ψ^2 is zero for a particular position and time, you would never find a particle there. Larger Ψ^2 values indicate higher probabilities of finding the particle at that particular position and time, as indicated in the sample wave function graph below.

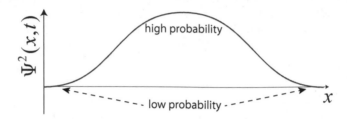

Going a step further, the **Heisenberg Uncertainty Principle** states that you cannot measure both an object's position and its momentum absolutely at the same time. The very act of attempting to measure one induces uncertainty in the other. Put another way, there will always be some amount of uncertainty in the measurement of an object's position and momentum, and the more exactly you know one measurement, the more uncertainty you have in the other.

9.27 Q: The wave function of a particle as a function of time is shown below. Rank the probability of finding the particle at the specified locations from highest to lowest.

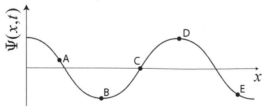

9.27 A: B=D, E, A, C. The square of the wave function indicates the probability of finding the particle at a given location and time.

The quantum mechanical model is consistent with the probabilistic description of matter. It's actually possible for electrons to experience spontaneous transitions to higher energy states, though it happens with a very low probability. On the other hand, spontaneous transitions to lower energy states occur with a very high probability. Regardless of the transition, the conservation laws continue to hold true.

Fission and Fusion

Recall that the nucleus of an atom is made up of protons and neutrons. Protons have a charge of +1 elementary charge, while neutrons are neutral. These nuclear particles are held together by the **strong nuclear force**. The **atomic number** of an atom (Z) describes the number of protons in the nucleus, while the **mass number** of an atom (A) describes the number of protons and neutrons in the nucleus. In a neutral atom, the number of orbiting electrons is equal to the number of protons. Atoms that are not neutral and have a non-zero net charge are known as **ions**.

The atomic number of an atom determines the element. A single element can have variations in its mass number, however, due to differing numbers of neutrons. The different configurations of an element having varying mass numbers are known as **isotopes** of that element, and are represented as:

$$_{Z}^{A}X^{charge}$$

9.32 Q: For the following atoms, determine the number of protons, neutrons, and electrons present:

A) $_{22}^{48}Ti^{+2}$

B) $_{35}^{77}Br$

C) $_{92}^{235}U^{-1}$

9.32 A: A) 22 protons, 26 neutrons, 20 electrons
B) 35 protons, 42 neutrons, 35 electrons
C) 92 protons, 143 neutrons, 93 electrons

In 1905, in a paper titled "Does the Inertia of a Body Depend Upon Its Energy Content," Albert Einstein proposed the revolutionary concept that an object's mass is a measure of how much energy that object contains, opening a door to a host of world-changing developments, eventually leading us to the major understanding that the source of all energy in the universe is, ultimately, the conversion of mass into energy!

If mass is a measure of an object's energy, we need to re-evaluate our statements of the law of conservation of mass and the law of conservation of energy. Up to this point, we have thought of these as separate statements of fact in the universe. Based on Einstein's discovery, however, mass and energy are two concepts effectively describing the same thing; therefore, we could more appropriately combine these two laws into a single law: the law of conservation of mass-energy. This law states that mass-energy cannot be created nor destroyed.

The concept of mass-energy is one that is often misunderstood and oftentimes argued in terms of semantics. For example, a popular argument states that the concept of mass-energy equivalence means that mass can be converted to energy, and energy can be converted to mass. Many would disagree that this can occur, countering that since mass and energy are effectively the same thing, you can't convert one to the other. For our purposes, we'll save these arguments for future courses of study. Instead, we will focus on a basic conceptual understanding.

The universal conservation laws studied so far include:

- Conservation of Mass-Energy
- Conservation of Charge
- Conservation of Linear Momentum
- Conservation of Angular Momentum

Einstein's famous formula, E=mc², relates the amount of energy contained in matter to the mass times the speed of light in a vacuum (c=3×10⁸ m/s) squared.

Theoretically, you could determine the amount of energy represented by 1 kilogram of matter as follows:

9.33 Q: What is the energy equivalent of 1 kilogram of matter?

9.33 A: $E = mc^2 = (1kg)(3 \times 10^8 \, \text{m/}_\text{s})^2 = 9 \times 10^{16} \, J$

This is a very large amount of energy. To put it in perspective, the energy equivalent of a large pickup truck is in the same order of magnitude of the total annual energy consumption of the United States!

More practically, however, it is not realistic to convert large quantities of mass completely into energy. Current practice revolves around converting small amounts of mass into energy in nuclear processes. Typically these masses are so small that measuring in units of kilograms is cumbersome. Instead, scientists often work with the much smaller **universal mass unit** (u), which is equal in mass to one-twelfth the mass of a single atom of Carbon-12. The mass of a proton and neutron, therefore, is close to 1u, and the mass of an electron is close to 5×10^{-4}u. In precise terms, $1u = 1.66053886 \times 10^{-27}$ kg.

One universal mass unit (1u) completely converted to energy is equivalent to 931 MeV. Because mass and energy are different forms of the same thing, this could even be considered a unit conversion problem.

The nucleus of an atom consists of positively charged protons and neutral neutrons. Collectively, these nuclear particles are known as nucleons. Protons repel each other electrically, so why doesn't the nucleus fly apart? There is another force which binds nucleons together, known as the **strong nuclear force**. This extremely strong force overcomes the electrical repulsion of the protons, but it is only effective over very small distances.

Because nucleons are held together by the strong nuclear force, you must add energy to the system to break apart the nucleus. The energy required to break apart the nucleus is known as the **binding energy** of the nucleus. This energy actually comes from a fraction of the mass of the nucleons themselves!

If measured carefully, you find that the mass of a stable nucleus is actually slightly less than the mass of its individual component nucleons. The difference in mass between the entire nucleus and the sum of its component parts is known as the **mass defect** (Δm). The binding energy of the nucleus, therefore, must be the energy equivalent of the mass defect due to the law of conservation of mass-energy: $E_{binding} = \Delta mc^2$.

Fission is the process in which a nucleus splits into two or more nuclei. For heavy (larger) nuclei such as Uranium-235, the mass of the original nucleus is greater than the sum of the mass of the fission products. Where did this mass go? It is released as energy! A commonly used fission reaction involves shooting a neutron at an atom of Uranium-235, which briefly becomes Uranium-236, an unstable isotope. The Uranium-236 atom then fissions into a Barium-141 atom and a Krypton-92 atom, releasing its excess energy while also sending out three more neutrons to continue a chain reaction! This process is responsible for our nuclear power plants, and is also the basis (in an uncontrolled reaction) of atomic fission bombs.

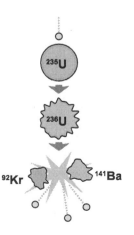

Fusion, on the other hand, is the process of combining two or more smaller nuclei into a larger nucleus. If this occurs with small nuclei, the product of the reaction may have a smaller mass its precursors, thereby releasing energy as part of the reaction. This is the basic nuclear reaction that fuels our sun and the stars as hydrogen atoms combine to form helium. This is also the basis of atomic hydrogen bombs.

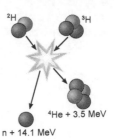

^{2}H ^{3}H

^{4}He + 3.5 MeV

n + 14.1 MeV

Nuclear fusion holds tremendous potential as a clean source of power with widely available source material (you can create hydrogen from water). The most promising fusion reaction for controlled energy production fuses two isotopes of hydrogen known as deuterium and tritium to form a helium nucleus and a neutron, as well as an extra neutron, while releasing a considerable amount of energy. Currently, creating a sustainable, controlled fusion reaction that outputs more energy than is required to start the reaction has not yet been demonstrated, but remains an area of focus for scientists and engineers.

9.34 Q: If a deuterium nucleus has a mass of 1.53×10^{-3} universal mass units (u) less than its components, what is its binding energy?

9.34 A: $(1.53 \times 10^{-3} u) \times \dfrac{9.31 \times 10^2 \, MeV}{1u} = 1.42 \, MeV$

9.35 Q: The energy produced by the complete conversion of 2×10^{-5} kilograms of mass into energy is

(A) 1.8 TJ

(B) 6.0 GJ

(C) 1.8 MJ

(D) 6.0 kJ

9.35 A: (A) $E = mc^2 = (2 \times 10^{-5} kg)(3 \times 10^8 \, {}^m\!/_s)^2 = 1.8 \times 10^{12} J = 1.8 TJ$

9.36 Q: A tritium nucleus is formed by combining two neutrons and a proton. The mass of this nucleus is 9.106×10^{-3} universal mass unit less than the combined mass of the particles from which it is formed. Approximately how much energy is released when this nucleus is formed?

(A) 8.48×10^{-2} MeV

(B) 2.73 MeV

(C) 8.48 MeV

(D) 273 MeV

9.36 A: (C) $9.106 \times 10^{-3} u \times \dfrac{9.31 \times 10^2\ MeV}{1u} = 8.48\ MeV$

9.37 Q: The energy equivalent of 5×10^{-3} kilogram is
(A) 8.0×10^5J
(B) 1.5×10^6J
(C) 4.5×10^{14} J
(D) 3.0×10^{19} J

9.37 A: (C) $E = mc^2 = (5 \times 10^{-3} kg)(3 \times 10^8\ {}^m\!/_s)^2 = 4.5 \times 10^{14} J$

Nuclear Decay

In the early 1900s, scientists began investigating radioactive materials. These materials release energy and/or particles when an unstable nucleus decays through one of three different processes. In the case of all these processes, the conservation laws continue to apply, but you can also incorporate a new conservation law specific to radioactive decay, the law of **conservation of nucleon number**.

The law of conservation of nucleon number states that the total number of particles in the nucleus, known as **nucleons**, must remain constant in any radioactive decay process. This conservation law, coupled with the law of conservation of electric charge, allows you to make predictions about nuclear reactions and nuclear decay.

In an **alpha decay** process, a particle consisting of a helium nucleus, comprised of two protons and two neutrons, also known as an **alpha particle** (α), is emitted, converting the initial element to a new element that has an atomic number two units less than the initial element.

$$\underset{Z}{\overset{A}{}}X \rightarrow \underset{2}{\overset{4}{}}He + \underset{Z-2}{\overset{A-4}{}}X'$$

In a **beta decay** process (β), a neutron decays into a proton (which remains in the nucleus) and an electron which is emitted. There are actually two types of beta decay. When an electron is emitted, this is known as β⁻ decay. It is also possible for beta decay to occur by emission of an anti-electron, also called a **positron**, in a process known as β⁺ decay. A positron has the same mass as an electron, but the opposite charge. An electron and a positron can annihilate each other, turning completely to energy as they release two photons traveling in opposite directions, each with an energy of 0.511 MeV.

$$\beta^- \text{ Decay: } {}^{A}_{Z}X \rightarrow {}^{0}_{-1}e + {}^{A}_{Z+1}X'$$

$$\beta^+ \text{ Decay: } {}^{A}_{Z}X \rightarrow {}^{0}_{1}e + {}^{A}_{Z-1}X'$$

In a **gamma decay** process (γ), a high-energy gamma photon is emitted. In this case, the number of protons and neutrons in the nucleus of the atom remains constant, but a gamma photon is emitted in a process involving rearrangement of the nucleons within the nucleus to a lower energy state that is analogous to the process by which an electron falls to a lower energy state.

$$_{Z}^{A}X \rightarrow \gamma + _{Z}^{A}X$$

An illustration summarizing the basic types of radioactive decay processes is shown below.

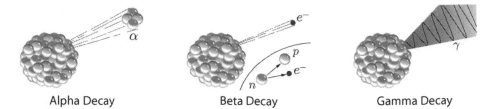

Alpha Decay   Beta Decay   Gamma Decay

These decay processes happen spontaneously to materials with unstable nuclei, known as radioactive materials. Though it is impossible to predict exactly when any single nucleus will undergo a nuclear decay process, you can predict what happens to a large number of identical nuclei using probability. The **half-life** of a material describes how long it takes for half of the nuclei to decay. Half-lives of materials can range from less than a billionth of a billionth of a second to billions of billions of years. Materials with shorter half-lives are much more unstable and decay more rapidly than materials with longer half-lives.

9.38 Q: After a uranium nucleus emits an alpha particle, the total mass of the new nucleus and the alpha particle is less than the mass of the original uranium nucleus. Explain what happens to the missing mass.

9.38 A: The missing mass is converted into energy.

9.39 Q: In the first nuclear reaction using a particle accelerator, accelerated protons bombarded lithium atoms, producing alpha particles and energy. The energy resulted from the conversion of mass into energy. The reaction can be written as shown below.

$$_1^1H + {_3^7}Li \rightarrow 2\,_2^4He + energy$$

Data Table

Particle	Symbol	Mass (u)
proton	$_1^1H$	1.007 83
lithium atom	$_3^7Li$	7.016 00
alpha particle	$_2^4He$	4.002 60

A) Determine the difference between the total mass of a proton plus a lithium atom, and the total mass of two alpha particles, in universal mass units.

B) Determine the energy in megaelectronvolts produced in the reaction of a proton with a lithium atom.

9.39 A: A) $(1.00783u + 7.01600u) - 2(4.00260u) = 0.01863u$

B) $0.01863u \times \dfrac{9.31 \times 10^2\,MeV}{1u} = 17.3\,MeV$

9.40 Q: The half-life of an imaginary element, physium, is approximately 60 days. If 100 kg is initially present, approximately how much remains after 240 days?

(A) 3 kg

(B) 6 kg

(C) 13 kg

(D) 25 kg

9.40 A: (B) 6 kg. After four half-lives, $(\frac{1}{2})^4$ remains.

9.41 Q: A radioactive hafnium sample with 72 protons and 106 neutrons per atom can undergo two reactions. For each reaction, describe the type of radioactive decay represented and determine the products of the reaction.

A) the nucleus emits a positron

B) the nucleus emits a gamma ray

9.41 A: A) Beta (β^+) Decay: $^{178}_{72}Hf \rightarrow\ ^{0}_{1}e + ^{178}_{71}Lu$

B) Gamma (γ) Decay: $^{178}_{72}Hf \rightarrow \gamma + ^{178}_{72}Hf$

9.42 Q: Determine what the ? represents in the following nuclear reactions and classify the associated nuclear decay process(es):

A) $^{230}_{91}Pa \rightarrow ? + ^{230}_{90}Th$

B) $^{238}_{92}U \rightarrow ? + ^{4}_{2}He + ^{234}_{90}Th$

C) $^{240}_{94}Pu \rightarrow ? + ^{236}_{92}U$

D) $^{228}_{88}Ra \rightarrow ? + ^{228}_{89}U$

9.42 A: A) $^{0}_{1}e$ Beta +
         B) γ Gamma and Alpha
         C) $^{4}_{2}He$ Alpha
         D) $^{0}_{-1}e$ Beta -

9.43 Q: The following equation is an example of what kind of nuclear reaction?

$$^{2}_{1}H + ^{2}_{1}H \rightarrow ^{3}_{2}He + ^{1}_{0}n + 3.27\,MeV$$

(A) fission
(B) fusion
(C) alpha decay
(D) beta decay

9.43 A: (B) fusion

9.44 Q: The mass m of a radioactive material was measured at the beginning of an experiement. Some time later, the amount of the original radioactive material was m/16. Determine the time between measurements if the half-life of the material is t.

(A) 2t
(B) 4t
(C) t²
(D) t/2

9.44 A: (B) 4t. The amount of radioactive material is halved four times, so the time of the experiment is four half-lives, or 4t.

9.45 Q: If a positron and an electron were completely converted to energy (annihilated) in a process releasing two photons traveling in opposite directions, which of the following must be true of the photons?

I. they must have the same wavelength

II. they must have the same energy

III. they must have the same momentum

(A) I and II only
(B) I and III only
(C) III only
(D) I, II, and III

9.45 A: (D) I, II, and III are all true. The photons must have equal momenta consistent with the law of conservation of linear momentum, therefore they also have the same wavelength, frequency, and energy.

9.46 Q: Five samples of radioactive materials with a given half-life are shown in the illustration below. Rank the mass of the remaining radioactive material after 2 years from greatest to least.

$m = 1kg$	$m = 2kg$	$m = 0.5kg$	$m = 2kg$	$m = 1kg$
$T_{\frac{1}{2}} = 1yr$	$T_{\frac{1}{2}} = 0.5yr$	$T_{\frac{1}{2}} = 0.5yr$	$T_{\frac{1}{2}} = 1yr$	$T_{\frac{1}{2}} = 0.5yr$

 A   B   C D E

9.46 A: D > A > B > E > C

Einstein's Relativity

Albert Einstein was puzzled by the concept of relative motion, especially when applied to situations of electromagnetic and gravitational forces. Einstein postulated that there are no instantaneous reactions in nature; therefore there must be a maximum possible speed for any reaction, which is the speed of light in a vacuum (c). Further, the speed of light in the vacuum must be the same for all observers, whether moving or at rest.

Einstein felt confident of these postulates, but soon realized that the repercussions of these postulates would upset the foundations of the known physical world. Eventually, he realized that the only way his postulates could be true in all situations involved a re-imagining of the concept of time, detailed in his proposal of special relativity. To begin with, events that appeared to occur simultaneously in one frame of reference did not necessarily occur simultaneously in another frame of reference. The entire concept of simultaneous events was relative!

More specifically, observers moving at different speeds would experience different time intervals. Imagine a modern training traveling at high speed. A laser, in the exact center of the train, turns on and shoots a laser beam toward the front and the back of the train. An observer in the center of the train car would see the laser beams hit the front and back windows of the train simultaneously. An observer on a lawn chair outside the train watching the train pass by, however, would see the laser exit the back window of the train first, since the back of the train is moving to meet the laser beam, while the front of the train is moving away from the laser beam. The observer outside the train does not see simultaneous events, while the observer inside the train does see simultaneous events. Time (and simultaneity) is therefore relative to the observer.

The implications of these findings are wide-spread and complex. Objects traveling at high velocities relative to an observer experience what is known as time dilation. What is experienced as a short time interval by the high-speed object is experienced as a longer time interval by the observer.

A famous thought experiment involves two identical twins on Earth. Suppose one twin leaves Earth at the age of 20 in an imaginary spaceship and travels at 90% of the speed of light (0.9c) a distance of 10 light years, then turns around and returns to Earth. The second twin remains on Earth. The space-faring twin experienced a trip of just under 20 years' time (under 20 years due to some other secondary relativistic effects), and has aged 20 years, returning as a 40-year-old, while the twin who remained on Earth experienced roughly 44 years while the sibling was in space, and is now 64 years old.

Further, as objects travel at higher speeds, their length contracts compared to the stationary observer. Further yet, as objects move faster and faster, it takes more and more energy to accelerate them further, therefore mass can never be accelerated to the speed of light.

Einstein generalized his work in the **Theory of General Relativity,** where he proposed that space and time are intertwined in a universal fabric known as spacetime. Large masses have the ability to bend the fabric of spacetime, leading to what you experience as gravity.

If all this sounds a bit complex and confusing, you're not alone. Entire courses and careers have been devoted to exploring and debating these theories. Thankfully, it is easy to find more resources on relativity, both on the web and in print form, written in a variety of formats from illustrated comics to complex mathematical proofs.

Most importantly, these developments serve as a great reminder that despite how much scientists think they know about the universe, there is much more that is yet unknown, and therefore so many more explorations to be undertaken and discoveries to be made that physics continues to grow and evolve on a daily basis.

9.47 Q: In a thought experiment, four objects are placed on a desk, at rest, and examined. Each object has the same radius and the given rest mass, as shown in the diagram below.

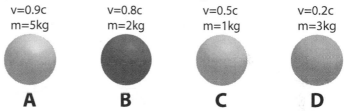

v=0.9c	v=0.8c	v=0.5c	v=0.2c
m=5kg	m=2kg	m=1kg	m=3kg
A	**B**	**C**	**D**

The objects are then accelerated to the right to the speeds given in the diagram. Rank the observed width of the moving objects from greatest to least from the reference point of a stationary observer.

9.47 A: D > C > B > A. The faster the objects go, the greater the length contraction effect. Note that accelerating such large masses to such extreme speeds would take tremendous amounts of energy, making actual performance of such an experiment impractical.

Test Your Understanding

1. Make a table comparing the various models of the atom discussed in this chapter. Highlight the strengths and weaknesses of each.

2. Why does UV light cause sunburn and lead to increased risk of cancer, yet visible light is considered safe?

3. How do scientists determine the composition of distant stars?

4. Use PhET's online simulation "Photoelectric Effect" to gather data on the element platinum. How can you use this simulation to determine the threshold frequency and work function of platinum? How can you use it to determine Planck's Constant? Do so.

5. An electron and a marble both travel through space at the same speed. Compare the wavelengths of the objects.

6. Play with the online PhET simulations "Alpha Decay," "Beta Decay," and "Nuclear Fission." How are these processes similar? How are they different?

7. Create a table comparing and contrasting alpha, beta, and gamma decay.

8. Can helium undergo alpha decay? Explain.

9. A container of hazardous radioactive waste has a half-life of 10 years. How could you determine how long this waste will remain hazardous?

10. Create a poster explaining the process of radioactive carbon dating.

11. If you piloted a futuristic spaceship toward the sun at a speed of 0.5c, at what speed would the light from the sun pass your ship?

12. Explain why an electron can never reach the speed of light.

Appendix A: AP-Style Problems

Fluids

1. A cart full of water travels horizontally on a frictionless track with initial velocity **v**. As shown in the diagram, in the back wall of the cart there is a small opening near the bottom of the wall that allows water to stream out. Considering just the cart itself (and not the water inside it), which of the following most accurately describes the characteristics of the cart?

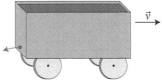

	Speed	**Kinetic Energy**
(A)	stays the same	stays the same
(B)	increases	increases
(C)	stays the same	increases
(D)	increases	stays the same

Answer: B

As the water streams out of the cart, the water is pushed out of the cart by pressure from above, exerting a reactionary force pushing the cart and its contents forward, increasing the speed of the cart. As the speed of the cart is increasing, and its mass remains the same, the kinetic energy of the cart must also increase.

---

2. A student places several ice cubes in a glass and fills the glass with water. After the ice cubes melt, the water level in the glass will

(A) rise a small amount

(B) fall a small amount

(C) remain at the same level

(D) not enough information given

Answer: C

The weight of the water displaced is equal to the weight of the ice cubes. Put another way, the volume of ice floating above the surface of the water matches the amount by which the water expanded when it turned to ice. Therefore, as the ice melts and returns to a liquid state, the water level remains constant.

3. A 20-kg uniform solid cylinder, A, is suspended underwater by a light rope. A second cylinder, B, with the same external dimensions but made of a higher density material is hollowed out and filled with pressurized helium gas such that, once sealed, it has the same total mass as the first cylinder. Cylinder B is then suspended underwater at the same depth by an identical rope. Compare the tension in the two ropes.

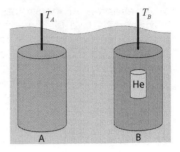

(A) $T_A < T_B$

(B) $T_A = T_B$

(C) $T_A > T_B$

(D) Not enough information given

Answer: B

Because the two cylinders have the same total mass encased in the same total volume, they have the same average density, and therefore, the same upward buoyant force acting upon them. The force of gravity on each cylinder is also the same as they have the same mass. Therefore, the same net force acts on each cylinder, and each rope must maintain the same tension.

4. A reservoir of incompressible fluid is topped by two pistons of non-equal areas, as shown in the diagram at right (not drawn to scale). The larger piston has four times the surface area of the smaller piston. If a force F pushes the smaller piston down by some distance d, which of the following best describes the upward force and distance traveled for the larger piston?

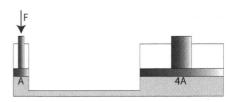

	Force	Distance Traveled
(A)	4F	d/4
(B)	F/4	4d
(C)	F	4d
(D)	4F	d

Answer: A

When a force is applied to a contained, incompressible fluid, the pressure increases equally in all directions throughout the fluid. Since the pressure at all points must be equal, the force per area must be equal; therefore the larger piston must have a four times larger force. Conservation of energy, however, dictates that the work done in pushing the small piston down must equal the work done in raising the larger piston; therefore, the larger piston only moves one fourth the distance of the smaller piston.

5. Two rain barrels, A and B, each contain the same volume of water at the same temperature, as shown at right. A pressure gauge is placed at the bottom of each barrel. Which statement best describes the pressures at the bottom of the barrels?

 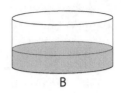

A   B

(A) $P_A > P_B$ since the pressure of the water depends upon the water's depth.

(B) $P_A < P_B$ since the pressure of the water depends upon the amount of atmosphere above the water.

(C) $P_A = P_B$ since the pressure of the water depends upon the volume of water.

(D) Not enough information given.

Answer: A

The pressure at the bottom of the barrels depends primarily upon the depth of the water.

---

6. Water flows through a section of thick piping with some velocity v as shown in the diagram at right. Based on the diagram, in which direction would you expect water to flow through the narrow section?

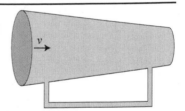

(A) to the right

(B) to the left

(C) remain stagnant

(D) not enough information given

Answer: A

Bernoulli's Principle states that fluids moving at higher velocities lead to lower pressures, and fluids moving at lower velocities result in higher pressures. Since there is a higher pressure at the left of the pipe, and a lower pressure at the right of the pipe, the water in the narrow section must flow from high to low pressure, or left to right.

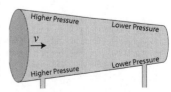

7. Water flows in a section of pipe of diameter d_1 with a constant volumetric flow rate Q_1 and linear velocity v_1. What happens to the volumetric flow rate Q_2 and linear velocity v_2 as the water transitions to a section of pipe with diameter d_2, which is one third the original pipe's diameter, as shown in the diagram?

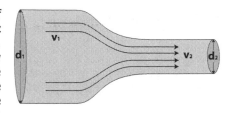

	Volumetric Flow Rate	Linear Velocity
(A)	Q/9	v
(B)	Q/3	9v
(C)	Q	9v
(D)	3Q	v/9

Answer: C

When fluids move through a full pipe, the volume of fluid that enters the pipe must equal the volume of fluid that leaves the pipe, even if the diameter of the pipe changes. This is a restatement of the law of conservation of mass for fluids. The volume flow rate, then, is equal to the area of the pipe multiplied by the velocity of the fluid. This must remain constant. Since the thin section of pipe has one-ninth the area of the thick section of pipe, the velocity of the fluid in the thin section must be nine times greater than in the thick section of pipe.

8. A vertical cylinder attached to the ground is partially evacuated to a pressure of 10,000 Pa by a vacuum pump. An airtight lid of radius 5 cm and mass 400 g is placed on its top. A rope is tied to the lid and a force of 800N vertically upward is applied. Which of the following best describes the result of the force applied to the rope?

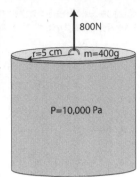

(A) The lid pops off the container and then accelerates upward at roughly 2000 m/s².

(B) The lid pops off the container and then accelerates upward at roughly 220 m/s².

(C) The lid pops off the container and then accelerates upward at roughly 0.23 m/s².

(D) The entire container, with sealed lid, remains stationary.

Answer: (A) The lid pops off the container and then accelerates upward at 2000 m/s². As the lid is pulled upward with a force of 800N, the force of gravity on the lid is roughly 4N down, and the force holding the lid of the cylinder to the body of the cylinder, determined from the difference in pressure on the inside and outside of the cylinder (90,000 pascals) multiplied by the area of the lid, is roughly 707N. This net imbalance in the force pops the lid off the cylinder, at which point it accelerates upward with a net force of 800N-4N = 796N upward. Applying Newton's 2nd Law, the acceleration of the lid is F/m or 796N/0.4kg = 2000 m/s².

Questions 9 and 10 refer to the following information: A square bedroom on the second floor of a house at sea level has a length of 4 meters.

9. What is the total downward force on the surface of the bedroom floor due to air pressure?

(A) 16 N

(B) 4×10^5 N

(C) 8×10^5 N

(D) 1.6×10^6 N

10. Three students examine the situation and make the following statements regarding why the force of the air pushing down on the floor does not collapse the floor:

Student A: Despite the large force due to air pressure on the surface of the floor, the floor is a high density solid and is attached to the much more massive Earth. The combined stability of the Earth and the flooring structure is more than enough to withstand the force of air pressure on the surface of the floor.

Student B: The force of the air underneath the floor pushing up very nearly balances the force of the air pushing down on the floor, creating a net force on the surface of the floor that is very nearly zero.

Student C: The force of the air pressure pushing down on the surface of the floor causes the floor to deform slightly, compressing the molecules of the floor, resulting in an elastic force pushing the air molecules back up. The elastic force of the floor balances the force of air pressure pushing down, resulting in a net force on the surface of the floor that is very nearly zero.

Which, if any, of these three students do you agree with and think is correct? Explain your reasoning.

Answers:

9. (D) 1.6×10^6 N $\qquad$ F=PA=(100,000 Pa)(4 m)^2 = 1.6×10^6 N

10. Student B - Pascal's Principle states that the force exerted by an enclosed fluid is applied to all surfaces of the container. Therefore, the force of air pressure in the room underneath the bedroom is exerting nearly the same force upward due to air pressure as the force of the air pushing down on the bedroom floor (neglecting the width of the floor itself, which makes a negligible difference in this problem).

---

11. A closed-tube mercury barometer has a height of 760 mm when placed in a room at one atmosphere (101,300 Pa) of pressure. If mercury has a density of 13,600 kg/m³, how tall would the barometer need to be if the liquid used inside it were honey (density 1400 kg/m³)?

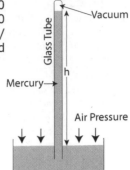

(A) 78 mm

(B) 7.4 m

(C) 78 m

(D) 7400 m

Answer: (B) 7.4 m. Utilizing Bernoulli's Equation you can solve for the height of the honey in the glass tube. Set side 1 as the air side, and side 2 as the inside of the tube.

$$P_1 + \tfrac{1}{2}\rho_1 v_1^2 + \rho_1 g h_1 = P_2 + \tfrac{1}{2}\rho_2 v_2^2 + \rho_2 g h_2 \xrightarrow[h_1=0, P_2=0]{v_1=v_2=0}$$

$$P_{air} = \rho_2 g h_2 \rightarrow h_2 = \frac{P_{air}}{\rho_2 g} = \frac{101,300 \, Pa}{(1400 \, ^{kg}\!/_{m^3})(9.8 \, ^m\!/_{s^2})} = 7.4m$$

12. A cylinder filled with water to a height h has two outlets. Water exits the barrel simultaneously through both outlet A and outlet B. Use this information and the diagram at right to answer the following question. Neglect friction.

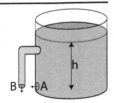

Where is the water velocity fastest?

(A) Fastest at A.

(B) Fastest at B.

(C) Fastest at whichever point has the smallest opening.

(D) They have the same velocity at A and B.

Answer: (D) They have the same velocity at A and B. Utilizing Bernoulli's Equation, both outlets are at the same depth from the surface, therefore they should have the same velocity.

Appendix A: AP-Style Problems

Thermal Physics

1. A canister of pressurized nitrogen gas at room temperature is cooled in a freezer. More nitrogen gas at room temperature is pumped into the canister until the canister returns to its initial pressure. The sealed canister is then returned to room temperature. A measurement of the pressure in the canister shows that the pressure in the canister is now twice its initial value.

Was more gas initially in the canister than was added?

(A) Yes

(B) No

(C) The amount of gas added to the canister is equal to the amount initially in the canister

(D) Not enough information is given

Answer: C

In order to answer this question, you only need to examine the initial and final conditions of the canister. Initially, the canister exhibited some pressure P. In its final condition, the gas has the same volume and temperature, but twice the pressure. If it has twice the pressure at the same volume and temperature, it must have twice the gas according to the ideal gas law PV=nRT. Therefore, the amount of gas in the canister was doubled, so the same amount of gas must have been added to the canister as was initially in the canister.

---

2. A refrigerator is placed in a room which is completely sealed except for a working electrical outlet. The refrigerator is plugged in, and the door of the refrigerator is opened. Does the temperature in the room increase, decrease, or remain the same? Justify your answer.

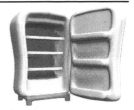

Answer: The temperature in the room increases. From an energy perspective, energy is entering the room through the electrical outlet, but it is not exiting the room. This energy accumulation will result in an increase in the room's temperature.

---

3. A power plant runs by means of a boiler used to vaporize water, turning a steam turbine which turns an electric generator, moving a coil of wire through a magnetic field to create electricity, as shown in the diagram.

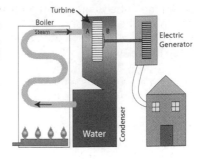

(a) Is the water vapor at position A (prior to striking the turbine) warmer than, cooler than, or the same temperature as the water vapor after it has struck the turbine at position B? Justify your answer.

(b) Assume you want to examine the efficiency of your power plant as a total system. Draw a dashed line encompassing the system you would examine.

(c) Is the amount of electrical energy delivered to the house greater than, less than, or the same as the amount of thermal energy generated by burning the fuel in the boiler? Justify your answer.

(d) Is the kinetic energy of the spinning turbine greater than, less than, or equal to the kinetic energy of the electric generator? Justify your answer.

(e) Is the average velocity of the water vapor molecules prior to hitting the turbine greater than, less than, or the same as the average velocity of the water vapor molecules after hitting the turbine? Justify your answer.

Answers:

(a) Warmer Than. As the water vapor molecules strike the turbine and transfer a portion of their momentum and energy to the turbine, they slow down, reducing their average temperature.

(b) Dashed line should encompass everything except the powerlines and house.

(c) Less Than. The system cannot be 100% efficient.

(d) More Information Needed. Even if you assume they rotate at the same speed (which may not be the case due to gearing assemblies), you do not know the rotational inertia of either the turbine or electric generator.

(e) Greater Than. The water vapor molecules slow down upon hitting the turbine, transfering a portion of their momentum and energy to the turbine.

4. An ice cube is placed in a cup of hot coffee. After a period of time, the ice cube melts. Which of the following best explains why the coffee becomes cooler? Select the two best answers.

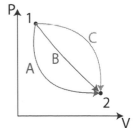

(A) The ice absorbs energy from the coffee as heat travels from the warmer to cooler material.

(B) The cool molecules in the ice radiate energy which is absorbed by the warmer molecules of the coffee.

(C) As the ice melts in the water, the cooler water molecules diffuse through the warmer coffee, lowering the average kinetic energy of the drink.

(D) The hot molecules in the coffee slow down as they are absorbed by the ice cube, causing the ice cube to melt.

Answers: (A) and (C). The ice absorbs energy from the coffee as heat travels from the warmer to cooler material. As the ice melts in the water, the cooler water molecules diffuse through the warmer coffee, lowering the average kinetic energy of the drink.

---

5. An ideal gas can move from state 1 to state 2 on a PV diagram by a variety of different pathways. Which of the following are the same regardless of pathway? Select two answers.

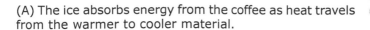

(A) The work done on the gas.

(B) The heat added to the gas.

(C) The change in average kinetic energy of the gas molecules.

(D) The change in the temperature of the gas.

Answers: (C) and (D). The change in average kinetic energy of the gas molecules and the change in the temperature of the gas. The temperature of the gas depends on the point on the PV diagram, and is independent of the path taken. As the temperature is related to the average kinetic energy of the molecules of the gas, the average kinetic energy of the gas molecules must also be path independent.

6. Students design an experiment to determine the thermal conductivity (k) of an unknown substance. A cube of the material to be tested with side area A is clamped to a steam chamber on one side, which maintains a constant temperature of 100°C, and a block of ice on the other side, which maintains a constant temperature of 0°C, for a difference in temperature from one side to the other of ΔT=100°C. The heat transferred is determined by finding the mass of water melted from ice, M_w, in time t.

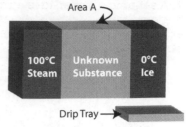

(a) Express the thermal conductivity of the substance as a function of the side area of the substance A, the latent heat of fusion of ice L_f, the mass of the melted water M_w, the time over which the water is collected t, and ΔT.

In ambient conditions, without the steam chamber attached, one gram of water is collected every 10 minutes. With the steam chamber attached, 5 grams of water are melted in 10 minutes. Ice has a latent heat of fusion (latent heat of melting) of L_f = 334 J/g. The unknown substance has a side length of 10 cm.

(b) Determine the thermal conductivity of the unknown substance.

Answers:

(a) $H = \dfrac{Q}{\Delta t} = \dfrac{kA\Delta T}{L} \xrightarrow[A=L^2 \rightarrow L=\sqrt{A}]{Q=M_w L_f} k = \dfrac{M_w L_f \sqrt{A}}{A\Delta t \Delta T} = \dfrac{M_w L_f}{\sqrt{A}\Delta t \Delta T}$

(b) $k = \dfrac{(4g)(334\,^J/_g)}{(0.1m)(600s)(100K)} = 0.223\,^J/_{s \cdot m \cdot K}$

7. A set amount of an ideal gas is taken through a thermodynamic cycle as shown in the PV diagram at right (not drawn to scale). Process DA is isothermal.

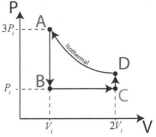

(a) Find the temperature at point C (T_C) in terms of the temperature at point B (T_B).
(b) Find the pressure at point D (P_D) in terms of P_i.
(c) Find the work done by the gas in moving from State A to State B to State C in terms of P_i and V_i.
(d) In which process or processes is heat taken from the gas? Explain.

Answers:

(a) The pressure at Point B is the same as the pressure at Point C, which is equal to P_i, therefore the ratio of the temperatures to the volumes at B and C must be equal:

$$P_B = P_C \rightarrow \frac{T_B}{V_B} = \frac{T_C}{V_C} \rightarrow T_C = \frac{V_C T_B}{V_B} = \frac{2V_i T_B}{V_i} = 2T_B$$

(b) Since Points A and D are on an isothermal line, they have the same temperature, so the product of their pressure and volume must be equal:

$$T_A = T_D \rightarrow P_A V_A = P_D V_D \rightarrow 3P_i V_i = P_D(2V_i) \rightarrow P_D = \tfrac{3}{2}P_i$$

(c) Work is the area under the graph, which is $P_i(2V_i - V_i) = P_i V_i$
(d) AB and DA. Examine each path separately:
   AB: $\Delta U = Q + W$, but $W=0$, so $\Delta U = Q$. Temperature decreases from A to B, so ΔU must be negative, therefore Q must be negative. This implies that heat is taken from the gas.
   BC: The gas is expanding so W is negative, and temperature is increasing, so ΔU is increasing while W is negative. Therefore, Q must be positive and heat is added to the gas.
   CD: $\Delta U = Q + W$, but $W=0$, so $\Delta U = Q$. Temperature increases from C to D so Q is positive, therefore heat is added to the gas.
   DA: ΔU is zero because DA is an isotherm, therefore $Q = -W$. The gas is compressed, so W is positive and Q must be negative, indicating that heat is taken from the gas.

8. An ideal gas moves from State A to State B following the path depicted in the PV diagram at right. How much work was done **on** the gas during this process?

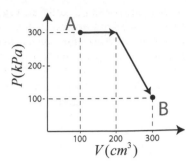

Answer: -50 J. The gas is expanding, therefore the gas is doing work. The answer to the question asked, then, how much work was done on the gas, must be negative. You can find the work done by taking the area under the graph, breaking the graph up into geometric shapes, as shown at right.

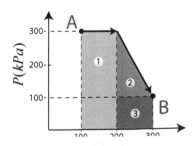

$A_1 = lw = (1 \times 10^{-4} m^3)(300{,}000\,Pa) = 30J$

$A_2 = \frac{1}{2}bh = \frac{1}{2}(1 \times 10^{-4} m^3)(200{,}000\,Pa) = 10J$

$A_3 = lw = (1 \times 10^{-4} m^3)(100{,}000\,Pa) = 10J$

$W_{total} = -(A_1 + A_2 + A_3) = -50J$

9. A cylinder filled with an ideal gas is fitted with a movable frictionless piston. Initially, the gas is in state A at 20 kPa, 400K, and 0.5 m³. The gas is taken through a reversible thermodynamic cycle as shown in the PV diagram at right (not drawn to scale).

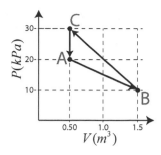

(a) Determine how many moles of gas are in the cylinder.
(b) Calculate the temperature of the gas at state B and state C.
(c) Determine the net work done on the gas during the cycle.
(d) Explain whether heat was added to the gas or removed from the gas during the cycle. Justify your answer.

Answers:

(a) At State A, T=400K, so you can use the ideal gas law to solve for the number of moles of gas:

$$PV = nRT \rightarrow n = \frac{PV}{RT} = \frac{(20,000\,Pa)(0.5m^3)}{(8.31 \text{\textfractionsolidus}_{mol \cdot K})(400K)} = 3.0 \; moles$$

(b) Knowing the quantity of the gas, these become straightforward ideal gas law applications:

$$T_B = \frac{P_B V_B}{nR} = \frac{(10,000\,Pa)(1.5m^3)}{(3.0mol)(8.31\text{\textfractionsolidus}_{mol \cdot K})} = 600K$$

$$T_C = \frac{P_C V_C}{nR} = \frac{(30,000\,Pa)(0.5m^3)}{(3.0mol)(8.31\text{\textfractionsolidus}_{mol \cdot K})} = 600K$$

(c) The net work done on the gas is the area enclosed by triangle ABC, and is positive as the gas cycle is counter-clockwise. You can determine the area of triangle ABC by finding the area of triangle BCD and subtracting the area of triangle ABD as shown at right.

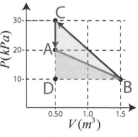

$$A_{BCD} - A_{ABD} = \tfrac{1}{2}(1m^3)(20,000\,Pa) - \tfrac{1}{2}(1m^3)(10,000\,Pa) = 5000J$$

(d) The work done on the gas is positive, and in a complete cycle ΔU=0, so Q=-W. Therefore, Q must have the opposite sign of W, and must be negative, indicating that heat was removed from the gas.

10. Absolute zero is the theoretical extrapolation of a volume-temperature plot for an ideal gas when the pressure becomes zero. Practically, you cannot reach absolute zero with an ideal gas because at very low temperatures, gases liquify. It is also challenging in a basic lab setting to adjust the volume of an ideal gas. However, by extrapolating the Pressure-Temperature plot of an ideal gas kept at a constant volume, you can arrive at a good approximation of absolute zero.

Students are presented with an enclosed sample of an ideal gas inside a closed 100 cm³ container.

   (a) What data would you need to collect in order to approximate absolute zero? Justify your answer.
   (b) Describe the equipment you would need to undertake this experiment. You may assume that probes can be placed into the container of gas without affecting the gas or its characteristics.
   (c) Design a plan for collecting this data.
   (d) Sketch an approximation of what you expect your best-fit line to look like on the axes at right. Indicate the portions of the graph where you anticipate gathering actual data, and portions of the graph where your line must be extrapolated.

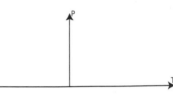

   (e) How can you determine the quantity of gas in the container from your graph?
   (f) How can you determine absolute zero from your graph?

Answers:

   (a) You would need to collect data on the pressure of the gas at varying temperatures in order to create the Pressure-Temperature plot.
   (b) A digital thermometer to measure the temperature of the ideal gas inside the container, and a pressure gauge to measure the pressure inside the container would be required. In addition, you would need a means of adjusting the temperature of the gas inside the container (heating / cooling equipment of some sort).
   (c) Starting at the minimum or maximum available temperature, read and record the pressure inside the container. Adjust the temperature and repeat until you reach the limits of the equipment and/or the gas liquifies.
   (d) See graph at right.
   (e) The slope of your line is equal to nR/V; therefore you can determine the number of moles of gas by rearranging for n such that n=slope×V/R.

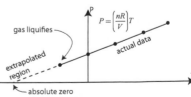

   (f) Absolute zero can be found at the x-intercept of your P-T plot as shown above.

Electrostatics

1. Three point charges are located at the corners of a right triangle as shown, where $q_1 = q_2 = 3\ \mu C$ and $q_3 = -4\ \mu C$. If q_1 and q_2 are each 1 cm from q_3, what is the magnitude of the net force on q_3?

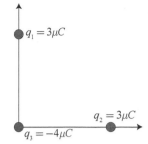

(A) 10.8 N

(B) 1080 N

(C) 1530 N

(D) 2160 N

Answer: (C) 1530 N

Besides utilizing Coulomb's Law correctly, the key to success in this problem is recognizing that the forces on q_3 are at right angles to each other. Therefore, to find the total, you must add them vectorially, finding their magnitude using the Pythagorean Theorem as they are at right angles to each other.

$$\left|\vec{F}_{1,3}\right| = \frac{kq_1q_2}{r^2} = \frac{(9\times10^9)(3\times10^{-6})(4\times10^{-6})}{(.01)^2} = 1080N \text{ up}$$

$$\left|\vec{F}_{2,3}\right| = \frac{kq_1q_2}{r^2} = \frac{(9\times10^9)(3\times10^{-6})(4\times10^{-6})}{(.01)^2} = 1080N \text{ right}$$

$$\left|\vec{F}_{total}\right| = \sqrt{(1080)^2 + (1080)^2} = 1530N \text{ up and to the right}$$

2. Three metal spheres rest on insulating stands as shown in the diagram at right. Sphere Y is smaller than Spheres X and Z, which are the same size. Spheres Y and Z are initially neutral, and Sphere X is charged positively. Spheres Y and Z are placed in contact near Sphere X, though Sphere X does not touch the other spheres.

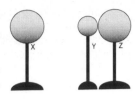

(a) Does sphere Y have a positive, negative, or neutral charge? Explain.
(b) Does sphere Z have a positive, negative, or neutral charge? Explain.
(c) Compare the magnitude of the charge on Spheres Y and Z.
(d) What happens to the magnitude of the charge on Sphere X? Explain.

Answers:

(a) Y has a negative charge. The force of attraction from Sphere X pulls electrons from Sphere Z into Sphere Y, giving it a net negative charge.
(b) Z has a positive charge. Z gives up excess electrons to Sphere Y, leaving it with a net positive charge.
(c) The magnitude of the charge on Spheres Y and Z is equal due to conservation of charge.
(d) The magnitude of the charge on Sphere X decreases due to a net flow of electrons from Y to X.

---

3. A capacitor is formed from two identical conducting parallel plates separated by a distance d. One plate is charged +Q, the other plate is charged -Q. A dielectric slab fills the space between the two plates. Where is the energy stored in this capacitor?

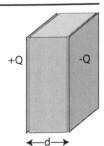

(A) On the outsides of the metal plates.

(B) On the insides of the metal plates.

(C) On the outside surface of the dielectric slab.

(D) Inside the dielectric slab.

Answer: (D) Inside the dielectric slab. The energy stored in a capacitor is actually stored in the form of electric field between the charged plates.

---

4. An air-gap parallel plate capacitor is attached to a source of constant potential difference as shown in the diagram at right. Which of the following statements is true if a dielectric is inserted between the plates?

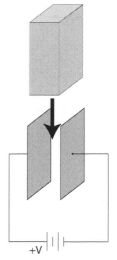

(A) Energy stored in the capacitor is reduced.

(B) Work must be done to insert the dielectric.

(C) Capacitance of the device is reduced.

(D) Electric field between the plates increases.

Answer: (B) Work must be done to insert the dielectric.

The insertion of the dielectric increases the capacitance of the device. The potential difference between the plates remains constant due to the voltage source; therefore, the electric field between the plates must remain constant as reflected by E=V/d. The energy stored in the capacitor must therefore increase, as reflected by the formula U=0.5CV². The additional energy comes from the work done to insert the dielectric.

5. Three point charges are situated as shown in the diagram at right.

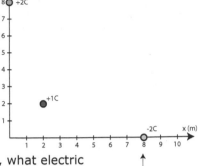

(a) Find the magnitude of the electric field at the origin due to the three charges shown.
(b) On the diagram below right, draw an arrow to represent the net electric field vector at the origin. Label angle theta in your diagram.
(c) Determine the angle theta in your diagram, in degrees.
(d) Calculate the electric potential at the origin.
(e) If an electron is placed at the origin, what electric potential energy does it possess? Answer in units of electron-volts.

Answers:

(a) First find the electric field due to each of the individual charges in terms of their vector components, then add those vectors to find the total electric field vector. You can then find the magnitude of the total electric field due to the three charges.

$$\vec{E}_{(0,8)} = \frac{kq}{r^2} = -\frac{(9 \times 10^9)(2)}{8^2}\hat{j} = -2.81 \times 10^8 \, \hat{j} \, {}^N\!/_C$$

$$\vec{E}_{(8,0)} = \frac{kq}{r^2} = -\frac{(9 \times 10^9)(2)}{8^2}\hat{i} = 2.81 \times 10^8 \, \hat{i} \, {}^N\!/_C$$

$$\left|\vec{E}_{(2,2)}\right| = \frac{kq}{r^2} = -\frac{(9 \times 10^9)(1)}{\left(\sqrt{2^2+2^2}\right)^2} = 1.13 \times 10^9 \, {}^N\!/_C \rightarrow \vec{E}_{(2,2)} = -7.95 \times 10^8 \, \hat{i} \, {}^N\!/_C - 7.95 \times 10^8 \, \hat{j} \, {}^N\!/_C$$

$$\vec{E}_{Total} = -5.14 \times 10^8 \, \hat{i} \, {}^N\!/_C - 1.08 \times 10^9 \, \hat{j} \, {}^N\!/_C$$

$$\left|\vec{E}_{Total}\right| = \sqrt{\left(-5.14 \times 10^8\right)^2 + \left(-1.08 \times 10^9\right)^2} = 1.20 \times 10^9 \, {}^N\!/_C$$

(b) See diagram at right.

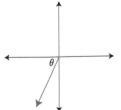

(c) $\theta = \tan^{-1}\left(\frac{opp}{adj}\right) = \tan^{-1}\left|\frac{1.08 \times 10^9}{5.14 \times 10^8}\right| = 65°$

(d) $V = \sum \frac{kq}{r} = \frac{(9 \times 10^9)(2)}{8} + \frac{(9 \times 10^9)(-2)}{8} + \frac{(9 \times 10^9)(1)}{\sqrt{2^2+2^2}} = 3.18 \times 10^9 V$

(e) $U = qV = (-1.6 \times 10^{-19} C)(3.18 \times 10^9 V) = -5.1 \times 10^{-10} J$

$-5.1 \times 10^{-10} J \times \frac{1eV}{1.6 \times 10^{-19} J} = -3.18 \times 10^9 eV$

6. A student investigates how a negatively charged ebonite rod influences an uncharged electroscope as shown in the diagram at right.

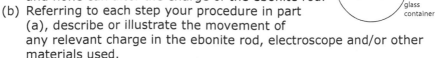

(a) Write a step-by-step procedure of how the negatively charged ebonite rod could produce a positive charge on the electroscope. Other materials can be used, but none can be charged, and none can alter the charge of the ebonite rod.
(b) Referring to each step your procedure in part (a), describe or illustrate the movement of any relevant charge in the ebonite rod, electroscope and/or other materials used.
(c) Assume the student charges the electroscope. Describe how the negatively charged ebonite rod could be used to test whether the electroscope is positively charged. Also explain what the student should expect to observe and why.

Answers:

(a) 1. Bring ebonite rod near, but do not touch, the knob of the electroscope.
2. While keeping rod near, ground the electroscope with a finger or grounding wire.
3. While keeping rod near, remove finger or grounding wire.
4. Remove ebonite rod.
(b) 1. No change in ebonite rod, electrons repelled to the leaves of the electroscope.
2. Electrons exit the electroscope into the ground/earth.
3. No charge movement in either device.
4. Positive and negative charges in electroscope redistribute such that the electroscope has a net positive charge.
(c) Bring the negatively charged rod near the electroscope; look for attraction between the leaves; the positively charged leaves are attracted to the opposite charge.

7. Two charged, parallel plates are separated by a distance, d. The electric field intensity, E, is 150 N/C. An electron with a speed of 3.0 × 10⁶ m/s comes in from the left just above the negatively charged plate as shown below.

(a) Draw the field lines between the plates.
(b) If the plates are 10 cm in length, how far has the electron moved in the y-direction during the time it takes to travel across the plates?
(c) Sketch the trajectory of the electron as it moves though the plates and beyond.
(d) How much KE has the electron gained as it moves between the plates?
(e) The electron is replaced by a proton moving at the same speed. How do the following change?

crossing time	___ increases	___ decreases	___ remains the same
acceleration	___ increases	___ decreases	___ remains the same
ΔKE	___ increases	___ decreases	___ remains the same

Answers:

(a)

(b) $t_{cross} = \dfrac{d}{v} = \dfrac{0.1m}{3\times10^6 \, m/_s} = 3.3\times10^{-8}s$

$a_y = \dfrac{F_y}{m} = \dfrac{qE_y}{m} = \dfrac{(1.6\times10^{-19}C)(150 \, N/_C)}{9.11\times10^{-31}kg} = 2.63\times10^{13} \, m/_{s^2}$

$\Delta y = \tfrac{1}{2}at^2 = \tfrac{1}{2}(2.63\times10^{13} \, m/_{s^2})(3.3\times10^{-8}s)^2 = 0.014m$

(c) See diagram above
(d) $W = \Delta K = F\Delta y = qE_y\Delta y = (1.6\times10^{-19}C)(150 \, N/_C)(0.014m) = 3.4\times10^{-19} \, J$

(e) Crossing time remains the same because no forces are acting in the direction of motion. Acceleration decreases because the force remains the same and the mass increases. Change in kinetic energy decreases because Δy decreases because acceleration decreases.

8. A charged conducting solid sphere of radius R has a charge of +Q.

(a) Describe and draw the shape and direction of the electric field lines inside of R and beyond R.

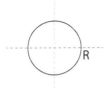

(b) Sketch a graph of electric field intensity as a function of distance from the center of the sphere, r.

(c) Draw three equipotential lines as dashed lines on the diagram below.

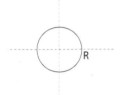

Answers:

(a) E=0 inside the sphere, so no lines. Field lines radiate outward in all directions starting at the surface of the sphere.

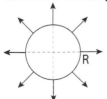

(b)    (c)

Circuits

1. Answer the following questions based on the scheµmatic at right, which shows a 3 µF and 6 µF capacitor connected in series, with a 2 µF capacitor connected in parallel to them. The system of capacitors is connected to a battery of voltage, V.

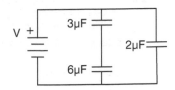

(a) Rank the potential differences across each capacitor and the battery. (1 indicates greatest potential. Give the same rank value for any that have the same potential difference.)

_____ $V_{2µF}$ _____ $V_{3µF}$ _____ $V_{6µF}$ _____ $V_{battery}$

(b) Show that the charge on the 3 µF capacitor must be the same value as the charge on the 2 µF capacitor.

(c) Calculate the ratio of the energy stored in the 2 µF capacitor to that of the 3 µF capacitor.

Answers:

(a) 1 $V_{battery}$; 1 $V_{2µF}$ because any capacitor in parallel has same potential as source (assuming no other caps in series w/source); 3 $V_{3µF}$ same charge as $V_{6µF}$ and since V=Q/C, lower C gives higher voltage drop; 4 $V_{6µF}$ higher C, lower V for given charge Q.

__1__$V_{2µF}$ __3__$V_{3µF}$ __4__$V_{6µF}$ __1__$V_{battery}$

(b) $Q_2=C_{2µF}V_{2µF}=(2µF)V_{battery}$ (since $C_{2µF}$ is in parallel w/battery). $Q_3=C_{3µF}V_{3µF}$ where $V_{3µF}=(2/3)V_{battery}$ ($V_{3µF}+V_{6µF}=V_{battery}$ and half the capacitance leads to twice the voltage drop), therefore $Q_{3µF}=3µF(2/3)V_{battery}=(2µF)V_{battery}$

(c) $U=Q^2/2C$, charge is the same for both, so $U_{2µF}/U_{3µF} = (1/C_{2µF})/(1/C_{3µF}) = 3/2$

     Appendix A: AP-Style Problems

2. A 2 µF, 3 µF, and 6 µF capacitor are connected in series to a 220-volt source. When completely charged, which of the following statements are true? Select two answers.

(A) The charge on the 2 µF capacitor is equal to 440 µC.

(B) The charge on the 3 µF capacitor is equal to 220 µC.

(C) The charge on all three capacitors is the same.

(D) The voltage drop across each of the three capacitors is equal to 220 V.

Answers: (B) and (C)

(A) is false because the voltage across the 2 µF capacitor is not 220 V, so $Q=CV=2$ µF × 220 V = 440 µC is false.

(B) is true since C_{eq} = 1 µF, so the total charge is $Q = CV = (1$ µF$)(220V) =$ 220 µC, and the charge on capacitors in series is the same.

(C) is true because the charge on capacitors in series is the same.

(D) is false because the sum of the voltage drops is 220V; therefore the voltage drop across each capacitor cannot be 220V.

---

3. Capacitors X, Y, and Z are connected in series to a voltage source and each has a capacitance of C, 2C, and 4C, respectively. When completely charged, which of the following choices gives the correct relationships among the stored energy in the capacitors?

(A) $U_X=2U_Y=4U_Z$

(B) $4U_X=2U_Y=U_Z$

(C) $U_X=U_Y=U_Z$

(D) $U_X=4U_Y=16U_Z$

Answer: (A) $U_X=2U_Y=4U_Z$ because in stored energy in capacitors is given by $U=Q^2/C$. Since the charges have to be the same value in a series configuration, the energy is inversely proportional to the capacitance.

4. Determine the effective capacitance of the circuit shown.

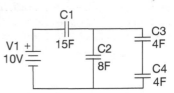

(A) 19 F

(B) 31 F

(C) 6 F

(D) (184/31) F

Answer: (C) 6 F. The two 4F capacitors in series reduce to 2F; added to the 8F capacitor in parallel gives 10F; 10F & 15F in series gives a total equivalent capacitance of 6F.

5. The figure below right shows a circuit with two batteries and three resistors, all labeled. Which of the following actions will increase the current through resistor R_2? (Select two answers.)

(A) Increasing V_1

(B) Increasing V_2

(C) Decreasing R_2

(D) Decreasing R_3

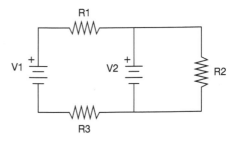

Answers: (B) & (C). The potential across R_2 is completely controlled by V_2. Therefore, to increase the current across R_2, you can either increase V_2 or decrease the resistance of R_2.

6. Capacitor X with a capacitance C is connected to a battery of voltage V. Capacitor Y of capacitance 2C is connected to another battery of voltage 4V. Both capacitors are then disconnected from the batteries and connected to each other in parallel.

   (a) What is the overall charge on both capacitors in terms of C and V?
   (b) What is the potential difference across the capacitors?
   (c) Using your answers from part (b), write an expression for the charge on each capacitor.
   (d) How does the energy stored in the capacitors before they were connected in parallel compare to the energy stored after they are connected in parallel? Justify your answer.

Answers:

   (a) Charge is conserved when the battery is disconnected, so when in parallel, $Q_{total}=Q_x+Q_y=CV+2C(4V)=9CV$

   (b) When disconnected, the charges will redistribute such that the potential across each is the same, but different than their original values. $V_x=V_y=Q_{total}/C_{eq}=(9CV)(C+2C)=3V$ (3 times the original voltage, V)

   (c) $Q_x=C_xV_x=C(3V)=3CV$ $Q_y=C_yV_y=2C(3V)=6CV$

   (d) $U_{after} < U_{before}$
   $$U_{before} = \tfrac{1}{2}CV^2 + \tfrac{1}{2}(2C)(4V)^2 = 16.5CV^2$$
   $$U_{after} = \tfrac{1}{2}C_{eq}(V')^2 = \tfrac{1}{2}(3C)(3V)^2 = 13.5CV^2$$

7. The circuits below depict identical batteries, resistors, and capacitors in various configurations. The circuits are initially open, and are all closed at the same time.

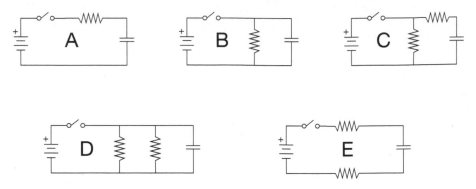

A) Rank the current through the battery immediately after the switch is closed from greatest to least.

B) Rank the current through the battery a long time after the switch is closed from greatest to least.

Answers: A) B=D, A, C, E; B) D, B=C, A=E. Note that when the switch is first closed, the capacitor acts like a wire. After a long time, the capacitor acts like an open.

8. The circuit depicted at right shows a battery, two identical resistors, and a capacitor. At time t=0 the switch is closed. The graphs below represent various circuit characteristics as a function of time.

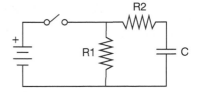

A

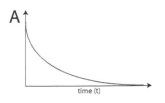

B

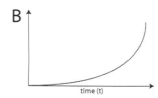

C

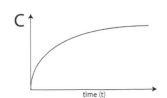

D

A) Which graph best represents the potential difference across R1?

B) Which graph best represents the current through R1?

C) Which graph best represents the potential difference across R2?

D) Which graph best represents the current through R2?

E) Which graph best represents the potential difference across the capacitor?

F) Which graph best represents the current flow in the capacitor?

Answers:

A) D

B) D

C) A

D) A

E) C

F) A

Magnetism

1. A charged particle is projected from point P with velocity v at a right angle to a uniform magnetic field directed out of the plane of the page as shown. The particle moves along a circle of radius R.

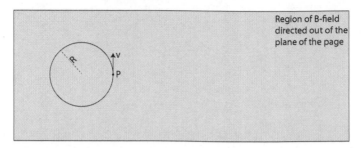

(a) On the diagram, draw a vector representing the magnetic force acting on the particle at point P.
(b) Determine the sign of the charge of the particle. Explain your reasoning.
(c) Explain why the magnetic field does no work on the particle as it moves in its circular path.
(d) A second, identically charged particle is projected at position P with a speed 2v in a direction opposite that of the first particle. On the diagram, draw the path followed by this particle. The drawn path should include a calculation of the radius of curature in terms of the original radius R.

Answers:

(a) Vector pointing toward center of circle.
(b) Negative. In order to apply the hand rule, the fingers of the left hand point in the direction of the particle's velocity, and bend out of the plane of the page, leaving the left-hand thumb pointing toward the center of the circle, creating the centripetal force allowing the particle to move in a circular path.
(c) No work is done because the magnetic force is always perpendicular to the velocity of the particle.
(d) See diagram at right. Radius of the new circle is 2R, since $mv^2/R=qvB$, therefore $R=mv/qB$.

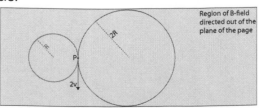

---

2. Radioactive sources emit alpha, beta and gamma radiation that enter the same uniform magnetic field at velocities measured in terms of the speed of light, c. Rank the magnitude of the magnetic force on each of the particles from lowest to highest for the following situations:

  (A) Alpha particle moving .1c at a right angle to the magnetic field.
  (B) Beta particle moving .2c, at a right to the magnetic field.
  (C) Gamma particle moving at c, at a right angle to magnetic field.
  (D) Alpha particle moving .1c, parallel to the magnetic field.
  (E) Beta particle moving .1c at a right angle to the magnetic field.

Answer: D=C < E < A=B

D & C experience no force. The force is zero for C since gamma particles are uncharged, and the force is zero for D since it is moving parallel to the magnetic field. E comes next since the beta particle has less charge than the alpha particle, and less speed than B. A and B come last and are equal since B has half the charge but twice the speed of A.

---

3. A uniform magnetic field is directed into the plane of the page. A loop of wire is placed in the magnetic field. At no time does the loop leave the magnetic field. Which of the following situations will induce a current in the loop? Select two answers.

  (A) rotate the loop along an axis that is directed into the page.
  (B) contract to loop to a smaller area.
  (C) rotate the loop along an axis that is directed vertically.
  (D) move the loop along a line that is parallel to the magnetic field.

Answers: B & C. A & D induce no current in the loop because there is no change in magnetic flux.

4. Four identical current-carrying wires are arranged at the corners of a square. A current, I, flows through each wire as shown. What is the direction of the net force on wire A due to wires B, C, and D?

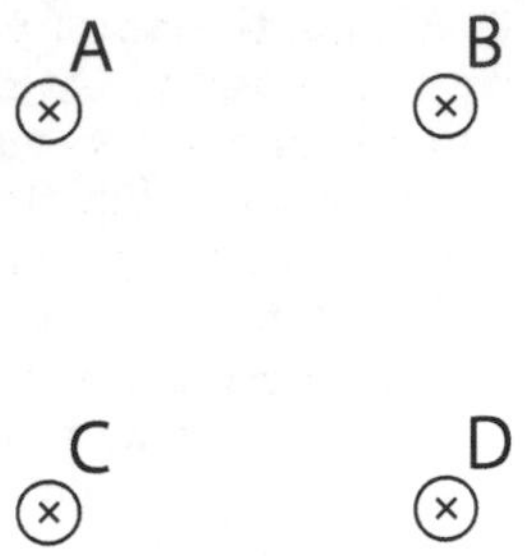

A ⊗   B ⊗

(A) Up
(B) Up and to the right
(C) Right
(D) Down and to the right
(E) Down

C ⊗   D ⊗

Answer: D. Wires with current flowing in the same direction attract each other, so the net force on A is down and to the right. You could also determine this by first finding the direction of the net magnetic field at A due to B, C, and D, then using the right-hand rule to determine the force on the charges flowing in wire A.

5. A square loop of wire with side length L and one side attached to an axis of rotation is situated in a uniform magnetic field directed into the page as shown. The magnetic field strength is B and a current I flows through the wire in a counter-clockwise direction.

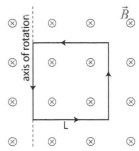

(a) Determine the net torque on the loop of wire.
(b) The magnetic field strength is now increased uniformly over a short period of time. Describe what happens to the current flowing in the wire during this period.
(c) The magnetic field is again set to strength B and is rotated 90 degrees such that it now points to the right, as shown. Determine the new net torque on the loop of wire.
(d) Again, the magnetic field strength is increased uniformly over a short period of time. Describe what happens to the current flowing in the wire during this period.

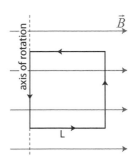

Answers:

(a) Net torque is zero as force on each segment of wire is toward the center of the loop.
(b) The current in the wire increases. An increasing magnetic flux through the loop creates a counter-clockwise current opposing the change in the magnetic flux. This induced current augments the initial current I.
(c) Only the segment of wire on the right side contributes to the net torque of the loop:
$$\tau = Fr = (BIL)r = BILr$$
(d) The flux through the loop doesn't change, so the current in the wire remains the same.

6. A particle of charge +q moves toward two long current-carrying wires at a right angle as shown in the diagram at right. If the distance between the wires is r, find the force on the particle when the particle is a distance r away from the nearest wire.

(A) $qV\dfrac{\mu_0 I}{4\pi r}$ up

(B) $qV\dfrac{\mu_0 I}{4\pi r}$ down

(C) $qV\dfrac{\mu_0 I}{2\pi r}$ up

(D) $qV\dfrac{\mu_0 I^2}{\pi r^2}$ down

Answer: B. First find the magnetic field strength at the point of interest by adding up the contributions to the magnetic field from each of the wires.

$$B_{net} = \frac{\mu_0 I}{2\pi r} \text{ out} + \frac{\mu_0 I}{4\pi r} \text{ in} = \frac{\mu_0 I}{4\pi r} \text{ out}$$

Next find the force on the charged particle, using the right-hand-rule to determine the direction as down (toward the bottom of the page).

$$\vec{F} = q\vec{v} \times \vec{B} \rightarrow |\vec{F}| = qvB\sin\theta \xrightarrow[\sin\theta=1]{\theta=90^\circ} |\vec{F}| = qv\frac{\mu_0 I}{4\pi r}$$

7. A magnetic field of strength B is oriented perpendicular to a rectangular loop of wire as shown. The width of the wire is w, and its length is L. The wire has a resistance per unit length of X ohms per meter. The wire is then uniformly rotated through an angle of θ about a vertical axis in a time period t. Assume θ is less than 90°.

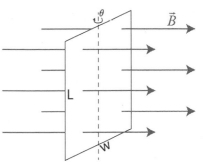

(a) Determine the resistance of the loop of wire.
(b) Determine the emf induced in the wire while the wire is rotating.
(c) Determine the induced current in the wire while the wire is rotating.
(d) Assume the given magnetic field strength B is uncertain. Describe a quick method / experiment to verify its accuracy. Specifically highlight any additional equipment you would require.

Answers:

(a) $R = X(2L+2w) = 2X(L+w)$

(b) $\varepsilon = \dfrac{\Delta\Phi_B}{\Delta t} = \dfrac{\Phi_{B_f} - \Phi_{B_i}}{t} = \dfrac{LwB\sin\theta - LwB}{t} = \dfrac{LwB}{t}(\sin\theta - 1)$

(c) $I = \dfrac{\varepsilon}{R} = \dfrac{LwB}{t}\dfrac{(\sin\theta - 1)}{2X(L+w)} = \dfrac{LwB(\sin\theta - 1)}{2Xt(L+w)}$

(d) There are a number of possible correct answers. One implementation could involve inserting an ammeter into the loop of wire (in series, of course) and running the exact scenario described, recording the angle through which the rectangle of wire is rotated, the time it takes to do so, and recording the ammeter reading. Magnetic field strength could then be determined from:
$$B = \dfrac{2IXt(L+w)}{Lw(\sin\theta - 1)}$$

8. A mass spectrometer is built such that an ion of known mass with charge +q is accelerated through a small slit such that it enters a velocity selector with speed v. The velocity selector is comprised a uniform electric field of strength E between two plates, placed in a region of uniform magnetic field strength B. A small slit at the end of the velocity selector admits particles that have passed through the region with no deflection into a region with just a uniform magnetic field, causing the ion to move in a circular path of radius r before striking a detector.

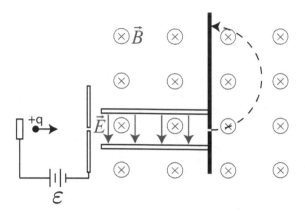

Answer parts (a) and (b) in terms of B, E, q, and r.

(a) Derive the speed of the ion such that it passes through the velocity selector undeflected.
(b) Determine the mass of the ion.

Following an equipment malfunction, it is observed that the radius of curvature of ions of the same type is 10% greater than it was when the machine was functioning correctly.

(c) Which of the following are possible causes of this discrepancy? Justify your answers.

_____ The emf of the battery has increased.

_____ The electric field strength has increased.

_____ The magnetic field strength has increased.

Answers:

(a) The electric force and the magnetic force must have matching magnitudes for the particle to pass through the velocity selector undeflected:

$$\left|F_e\right| = \left|F_B\right| \rightarrow qE = qvB \rightarrow v = \frac{E}{B}$$

(b) Determine the mass of the ion by recognizing that in the deflection chamber, the centripetal force is caused solely by the magnetic force on the moving charged particle:

$$F_c = qvB = \frac{mv^2}{r} \rightarrow qB = \frac{mv}{r} \rightarrow v = \frac{qrB}{m} \xrightarrow{v=\frac{E}{B}} \frac{E}{B} = \frac{qrB}{m} \rightarrow m = \frac{qrB^2}{E}$$

(c) _No__ The emf of the battery has increased. Increasing the emf of the battery would allow more energetic ions into the velocity selector, but only ions of the original velocity would exit the velocity selector.

Yes The electric field strength has increased. Increasing the electric field strength would allow ions with higher speeds into the deflection chamber, resulting in a larger radius of travel.

_No__ The magnetic field strength has increased. Increase the magnetic field strength would allow ions with lower speeds into the deflection chamber, resulting in a smaller radius of travel.

Optics

1. Light from an object passes through a converging lens and is focused to form an image as shown in the diagram at right. Which light ray reaches the image plane in the least amount of time?

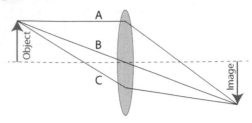

(A) Ray A

(B) Ray B

(C) Ray C

(D) They all take the same amount of time.

Answer: (D) They all take the same amount of time. Light traveling between any two points always takes the path that takes the least amount of time. Since light from the object reaches the image plane by all three paths, no one path can take any longer than the others, or light wouldn't travel that path.

---

2. Light of frequency 1.5×10^{14} Hz travels through air and enters glass perpendicular to its surface. Which of the following changes occur to the ray of light as it enters the glass? Select two answers.

(A) wavelength decreases

(B) speed of the ray of light decreases

(C) light ray bends toward the normal

(D) frequency of the ray of light increases

Answers: (A) and (B) are correct.

A is correct because wavelength is inversely proportional to the index of refraction. B is correct because the velocity is inversely proportional to the index of refraction. C is incorrect because no refraction occurs when the angle of incidence is 0 degrees. D is incorrect because frequency depends on the source of light.

---

3. A right angle prism with an index of refraction of n=1.84 is shown at right.

What is the maximum angle of incidence on the prism face such that total internal reflection does **not** occur on the back face of the prism?

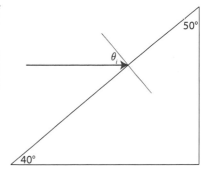

Answer: 24.4°

First find the critical angle for the back face:

$$\sin\theta_C = \frac{1}{n_{prism}} \rightarrow \theta_C = \sin^{-1}\left(\frac{1}{n_{prism}}\right) = \sin^{-1}\left(\frac{1}{1.84}\right) = 33°$$

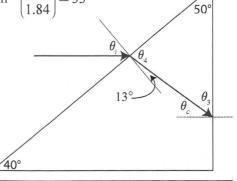

Next, note that a ray incident on this back face at this angle would make an angle of 57° with the back face (θ_3).

This would lead to a refracted angle of 13° at the prism face (θ_4=73°).

Applying Snell's Law, the incident angle is determined as 24.4°.

4. Parallel rays of monochromatic light are incident upon two identical diverging lenses in series before reaching a white sheet of paper. The intensity of the light upon the paper as measured at Point P is

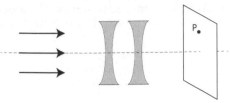

(A) greater in this configuration than with only a single lens.

(B) less in this configuration than with only a single lens.

(C) the same in this configuration than with only one lens.

(D) more information needed.

Answer: (B) Two diverging lens in series spread out the light more than a single diverging lens; therefore, the energy of the light incident upon the paper is spread out over a larger area with two diverging lenses, resulting in less intensity as measured at point P.

5. An electromagnetic wave travels through space. Which of the following best describe the orientation of the magnetic field? Select two answers.

(A) The magnetic field is parallel to the velocity of the wave.

(B) The magnetic field is perpendicular to the velocity of the wave.

(C) The magnetic field is parallel to the electric field.

(D) The magnetic field is perpendicular to the electric field.

Answers: (B) & (D). The magnetic field is oriented perpendicular to both the velocity of the wave and the electric field.

6. The diagram below shows plane wave fronts as they travel from medium 1 to medium 2 across the boundary AB. The wave fronts are drawn to scale.

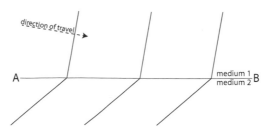

(a) Explain and justify how one could use a ruler to show that the speed in medium 2 is less than in medium 1.
(b) A normal line has been drawn at one of the intersections of the wave fronts in each medium and the boundary AB. Draw appropriate incident and refracted rays at this normal and label the angle of incidence (θ_i) and angle of refraction (θ_r).

(c) Explain and justify how one could use a protractor to show that speed of the wave in medium 2 is less than in medium 1.

Answers:

(a) Lay the ruler perpendicular to the wave fronts in each medium and measure the distance between two successive waves. This gives the wavelength of each. The wavelength in medium 2 is less than in medium 1, therefore its speed is less as well (wavelength is proportional to velocity according to the wave equation $v=f\,\lambda$.
(b) The rays should be perpendicular to the wave fronts and angles measured relative to the normal line as shown in the diagram below.

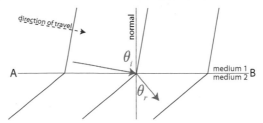

(c) Use a protractor to measure the angles of incidence and refraction. Since $\theta_r < \theta_i$, the speed in medium 2 is less than the speed in medium 1, consistent with Snell's Law.

7. Light of wavelength λ in a vacuum has what wavelength in a material with index of refraction n?

(A) $n\lambda$

(B) $\dfrac{n}{\lambda}$

(C) $\dfrac{\lambda}{n}$

(D) $\dfrac{n}{c\lambda}$

Answer: (C) $n = \dfrac{c}{v} \xrightarrow{v=f\lambda} n = \dfrac{c}{f\lambda_{new}} \rightarrow \lambda_{new} = \dfrac{c}{fn} \xrightarrow{c=f\lambda} \lambda_{new} = \dfrac{f\lambda}{fn} = \dfrac{\lambda}{n}$

8. Green monochromatic light of wavelength 550 nm passes through a diffraction grating of 2000 lines/cm. Determine the distance between the first and second bright spots on a screen 1 m from the grating.

(A) 5×10^{-6} m

(B) 0.115 m

(C) 0.226 m

(D) 0.639 m

Answer: (B) 0.115 m

First find the distance between the diffraction grating slits.

$$d = \frac{1}{2000 \, lines/_{cm}} = 5 \times 10^{-4} \, cm/_{line} \times \frac{1m}{100cm} = 5 \times 10^{-6} \, m/_{line}$$

Next, find the angle for the first order bright spot.

$$m\lambda = d \sin\theta \rightarrow \sin\theta = \frac{m\lambda}{d} \rightarrow \theta = \sin^{-1}\left(\frac{m\lambda}{d}\right) = \sin^{-1}\left(\frac{550 \times 10^{-9} \, m}{5 \times 10^{-6} \, m/_{line}}\right) = 6.32°$$

With this angle, you can find the vertical displacement of the first order bright spot.

$$\tan\theta = \frac{opp}{adj} \rightarrow opp = adj \tan\theta = 1m \tan(6.32°) = 0.111m$$

Then, find the angle for the second order bright spot.

$$\theta = \sin^{-1}\left(\frac{m\lambda}{d}\right) = \sin^{-1}\left(\frac{2 \times 550 \times 10^{-9} \, m}{5 \times 10^{-6} \, m/_{line}}\right) = 12.71°$$

With this angle, find the vertical displacement of the second order bright spot.

$$opp = adj \tan\theta = 1m \tan(12.71°) = 0.226m$$

The distance between the bright spots is therefore 0.226m-0.111m=0.115m

9. A concave spherical mirror with a 10-cm radius of curvature sits on a principal axis as shown. An object of height 6 cm is placed 15 cm to the left of the mirror.

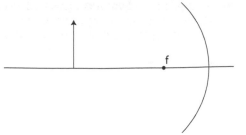

(a) On the diagram above, locate the image by ray tracing.
(b) Calculate the position of the image.
(c) Characterize the image as real or virtual, upright or inverted.
(d) Determine the height of the image.

Answers:

(a) See diagram below:

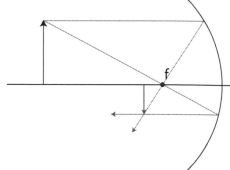

(b) First find the focal length of the mirror as half the radius:
f = R/2 = 10 cm/2 = 5 cm.
$$\frac{1}{f} = \frac{1}{d_o} + \frac{1}{d_i} \rightarrow \frac{1}{5cm} = \frac{1}{15cm} + \frac{1}{d_i} \rightarrow d_i = 7.5cm$$

(c) The image is inverted and real.

(d) $m = \dfrac{-d_i}{d_o} = \dfrac{h_i}{h_o} \rightarrow \dfrac{-7.5cm}{15cm} = \dfrac{h_i}{6cm} \rightarrow h_i = -3cm$

10. A sheet of crown glass (n=1.52) is coated with a thin transparent film of water (n=1.33). A beam of green monochromatic light of wavelength 500nm strikes the surface at an angle of 90°.

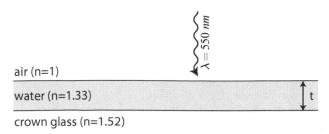

air (n=1)

water (n=1.33)

crown glass (n=1.52)

(a) What is the frequency of the light in the air?
(b) What is the frequency of the light in the thin film of water?
(c) What is the wavelength of the light in the thin film of water?
(d) What is the minimum thickness of water that will minimize reflection of this light?

Answers:

(a) $v = f\lambda \rightarrow f = \dfrac{v}{\lambda} = \dfrac{3\times10^8 \, m/s}{550\times10^{-9} m} = 5.45\times10^{14} \, Hz$

(b) Same as a.

(c) $\lambda_{film} = \dfrac{\lambda}{n} = \dfrac{550\times10^{-9} m}{1.33} = 414nm$

(d) For destructive interference, the total optical path difference should be half the wavelength of the light in the film. Since the light reflects off higher-index materials twice, an even number, there is no phase shift.

$OPD = 2t = \dfrac{\lambda_{film}}{2} \rightarrow t = \dfrac{\lambda_{film}}{4} = \dfrac{414nm}{4} = 103.5nm$

Modern Physics

1. A continuous white-light spectrum illuminates an imaginary monatomic gas with a ground state of -5.0 eV. An optical analyzer reports that wavelengths of 405 nm and 546 nm are absorbed by the gas.

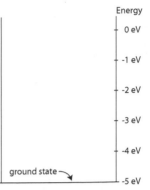

(a) Determine the energies of the photons of light absorbed by the gas.

(b) Show all the excited energy states for the gas atoms in the energy-level diagram above right.

(c) On the energy-level diagram, indicate all possible electron transitions that would produce bright lines in an emission spectrum by drawing arrows showing the transitions.

(d) What is the wavelength of the lowest energy photon possibly produced by an electron relaxing from a higher level energy state to a lower level energy state?

(e) What type of radiation does this wavelength represent?

Answers:

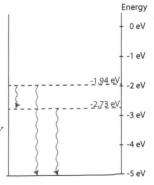

(a) $E = \dfrac{hc}{\lambda} = \dfrac{1240eV \bullet nm}{405nm} = 3.06eV$

$E = \dfrac{hc}{\lambda} = \dfrac{1240eV \bullet nm}{546nm} = 2.27eV$

(b) see diagram

(c) see diagram

(d) $E_{photon} = E_i - E_f = -1.94eV - -2.73eV = 0.79eV$

$E = \dfrac{hc}{\lambda} \rightarrow \lambda = \dfrac{hc}{E} = \dfrac{1240eV \bullet nm}{0.79eV} = 1570nm$

(e) Infrared

---

2. Which of the following types of nuclear decay particles have the same mass as an electron? (Select two answers.)

(A) Alpha particle

(B) Beta particle

(C) Positron

(D) Gamma Ray

Answer: (B) beta particle & (C) positron. A beta particle is an electron, and a positron is an anti-electron, a particle with the same mass as an electron but the opposite charge.

---

3. A specific radioactive material is found to decay such that 25% of the original sample remains after 100 years. What is the half-life of the material?

(A) 25 years

(B) 33.3 years

(C) 50 years

(D) 66.7 years

Answer: (C) 50 years. After one half-life (50 years), half the sample remains. After a second half life (another 50 years), one quarter (25%) of the sample remains.

---

4. For each of the following reaction equations, identify the type of nuclear reaction.

(A) $^{236}_{92}U \rightarrow \, ^{144}_{56}Ba + \, ^{89}_{36}Kr + 3\,^{1}_{0}n + 177\,MeV$

(B) $^{40}_{18}Ar^{*} \rightarrow \, ^{40}_{18}Ar + \gamma$

(C) $^{3}_{1}H + \, ^{2}_{1}H \rightarrow \, ^{4}_{2}He + \, ^{1}_{0}n + 17.6\,MeV$

(D) $^{40}_{19}K \rightarrow \, ^{40}_{18}Ar + \, ^{0}_{1}\beta$

Answers:

(A) fission

(B) gamma decay

(C) fusion

(D) β⁺ decay

5. Which of the following statements concerning Einstein's Theory of Special Relativity is true?

(A) Events occur simultaneously for all observers in all reference frames.

(B) The speed of light has the same value for all observers in all reference frames.

(C) A clock moving at a high rate of speed runs slower than a clock at rest.

(D) A car moving at a high rate of speed is longer than a car at rest.

Answer: (B) The speed of light has the same value for all observers in all reference frames. This is the foundation of special relativity.

6. The following fusion reaction occurs in stars:

$$^{21}_{10}Ne + {}^{4}_{2}He \rightarrow {}^{24}_{12}Mg + {}^{1}_{0}n$$

Given the masses listed in the table at right, how much energy is released in the reaction?

Particle	Mass (u)
$^{1}_{0}n$	1.007825
$^{4}_{2}He$	4.002603
$^{21}_{10}Ne$	20.993849
$^{24}_{12}Mg$	23.985042

Answer: 3.34 MeV. Add up the mass of all the reagents, add up the mass of all the products, and the difference is the amount of mass converted into energy. 1 universal mass unit (u) results in 931 MeV.

---

7. Which of the following best supports the particle nature of light?

(A) Compton Effect

(B) Diffraction

(C) Doppler Effect

(D) Interference

Answer: (A) The Compton Effect showed that photons have momentum and they obey the laws of conservation of energy and conservation of momentum.

8. An unknown metal is irradiated with monochromatic light in an experiment. As the frequency of the light is increased, electrons are ejected and collected in a photoelectric circuit. A reverse-bias potential is applied to the circuit at each specific frequency until the current just reaches zero. This stopping potential is recorded for the various incident frequencies and plotted below.

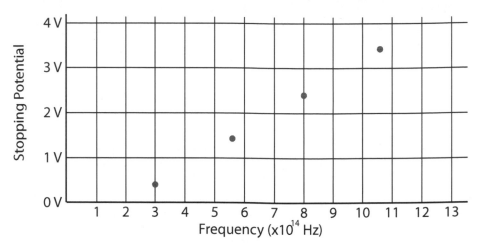

(a) What is the maximum speed an emitted photoelectron could obtain if this metal were irradiated with light of wavelength 429 nm?
(b) Determine the cutoff frequency of the metal.
(c) Determine the work function of the metal.
(d) Use the graph to calculate Planck's Constant. Compare to the accepted value and calculate a percent error.
(e) Draw a line representing a different metal that has a work function of 1.5 volts.

Answers:

(a) First find the energy of the photoelectron. To do this, determine its frequency and use the graph to find the stopping potential of the emitted particle:

$$v = f\lambda \rightarrow f = \frac{v}{\lambda} = \frac{3\times10^8 \,{}^m\!/_s}{429\times10^{-9}\,m} = 7\times10^{14}\,Hz$$

This stopping potential correlates to the energy of the emitted photoelectron. Use this energy to find the speed of the emitted electron.

$$2eV = 3.2\times10^{-19}\,J$$

$$KE = \tfrac{1}{2}mv^2 \rightarrow v = \sqrt{\frac{2\,KE}{m}} = \sqrt{\frac{2(3.2\times10^{-19}\,J)}{9.11\times10^{-31}\,kg}} = 838{,}000 \,{}^m\!/_s$$

(b) 2×10^{14} Hz right from the graph.
(c) You can extend the graph to find the y-intercept, which is the work function, or use the equation of the line:

$$m = slope = \frac{rise}{run} = \frac{4V - 0V}{12\times10^{14}\,Hz - 2\times10^{14}\,Hz} = 4\times10^{-15}\,V \bullet s$$

Next utilize the formula for a line, y=mx+b, to solve for the y-intercept.

$$y = mx + b \rightarrow b = y - mx \underset{m=4\times10^{-15}V\bullet s}{\overset{y-2V,x-7\times10^{14}\,Hz}{\longrightarrow}} b = 2V - (4\times10^{-15}V \bullet s)(7\times10^{14}\,Hz) = -0.8V$$

The work function is therefore 0.8V.
(d) The slope leads you to Planck's Constant, though you must treat the y-axis as the electron energy by multiplying the potential by the charge on an electron.

$$slope = \frac{rise}{run} = \frac{4eV \times 1.6\times10^{-19}\,{}^J\!/_{eV}}{10\times10^{14}\,Hz} = 6.4\times10^{-34}\,J \bullet s$$

Next find the error in your measurement:

$$\%err = \frac{\left|h_{exp} - h_{acc}\right|}{h_{acc}} \times 100\% \rightarrow$$

$$\%err = \frac{\left|6.4\times10^{-34}\,J\bullet s - 6.63\times10^{-34}\,J\bullet s\right|}{6.63\times10^{-34}\,J\bullet s} \times 100\%$$

$$\%err = 3.5\%$$

(e) This line will have the same slope as the previous line, but the y-intercept will be -1.5, so you can find two points on the line using the equation of the line y=mx+b, which in this case will be y=(4×10⁻¹⁵ V.s)x-1.5V. Example points could include (3.75×10¹⁴ Hz, 0) and (10×10¹⁴ Hz, 2.5V). Connect the points to get the line!

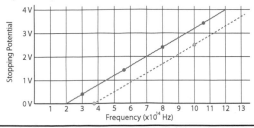

9. An experiment is set up such that a beam of monochromatic light passes through a diffraction grating with 3000 lines/cm, creating a diffraction pattern on a screen located 0.1 meter from the grating. The distance between the central maximum and the next nearest bright line on the screen is 1.32 cm.

(a) Determine the wavelength of the incident light.
(b) What is the energy of each incident photon (in eV)?

The incident light is created by a filtered mercury arc lamp. A few energy levels of mercury are shown at right.

(c) Which electron transition in the lamp is most likely responsible for creation of the photons striking the diffraction grating?
(d) The diffraction grating is replaced with a new grating containing 2000 lines/cm. Does the distance between maxima
___ increase
___ decrease
___ remain the same

Explain your reasoning.

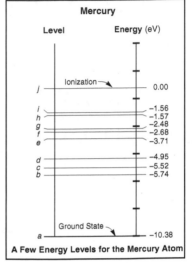

Mercury

Level | Energy (eV)

Ionization → ... 0.00

j

i ... –1.56
h ... –1.57
g ... –2.48
f ... –2.68
e ... –3.71

d ... –4.95
c ... –5.52
b ... –5.74

Ground State ...

a ... –10.38

A Few Energy Levels for the Mercury Atom

Answers:

(a) First find the spacing between slits: $\dfrac{.01m}{3000 lines} = 3.33 \times 10^{-6} m$

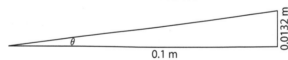

0.0132 m

0.1 m

Next determine a value for sin θ given your geometric setup:

$$\theta = \tan^{-1}\left|\frac{0.0132m}{0.1m}\right| = 7.52° \rightarrow \sin\theta = \sin(7.52°) = 0.1308$$

Then solve for the wavelength of light using the diffraction grating equation:

$$d\sin\theta = m\lambda \rightarrow \lambda = \frac{d\sin\theta}{m} \xrightarrow[\sin\theta=0.1308, m=1]{d=3.33\times10^{-6}m} \lambda = 4.36\times10^{-7}m = 436nm$$

(b) $E = hf = \dfrac{hc}{\lambda} = \dfrac{1240eV \bullet nm}{436nm} = 2.84eV$

(c) Level f to level c: $E_{photon} = E_i - E_f = -2.68eV - -5.52eV = 2.84eV$

(d) Decrease. With fewer slits per cm, the width of the slits increases. therefore, the angle must decrease.

10. An alien space ship attacks a planet by shooting a laser beam at the planet every five seconds. If the ship is approaching the planet at high speed, which of the following is true for an observer situated on the planet?

(A) The laser beams strike the planet at a frequency of 0.2 Hz.

(B) The laser beams strike the planet at a frequency less than 0.2 Hz.

(C) The laser beams strike the planet at a frequency greater than 0.2 Hz.

(D) More information is needed.

Answer: (C) The laser beams strike the planet at a frequency greater than 0.2 Hz. As the spaceship approaches the planet while emitting laser beams at a frequency of 0.2 Hz, the observer on the planet will observe a frequency shift to a higher frequency consistent with the Doppler Effect.

---

11. Barry and Dawn meet at 11:59 pm every New Year's Eve at the top of the Sears Tower in Chicago. This year, Barry, a pilot, travels 1,500,000 kilometers during the year, while Dawn, a computer programmer, travels 20,000 km during the year. For the period between 11:59 pm last year and 11:59 pm this year, which of the following is true?

(A) Barry has aged less than Dawn.

(B) Dawn has aged less than Barry.

(C) They have both aged the same amount.

(D) Not enough information is given.

Answer: (A) Barry has aged less than Dawn. According to relativity theory, the faster an object moves, the less time it experiences. Put another way, the two events at 11:59 pm are separated by a specific amount of space-time. Since Barry travels further in space, he must experience less time than Dawn, who travels less in space, and therefore experiences more time.

---

12. An X-ray photon with frequency f undergoes an elastic collision with a stationary electron (mass m_e) and is scattered. The frequency of the scattered X-ray photon is f'.

(a) Determine the kinetic energy of the electron after the collision in terms of f, f', and fundamental constants.
(b) Determine the magnitude of the momentum of the scattered electron in terms of f, f', m_e, and fundamental constants.
(c) Determine the de Broglie wavelength of the electron in terms of f, f', m_e, and fundamental constants.

Answers:

(a) Utilizing conservation of energy:
$$hf = K_{electron} + hf' \rightarrow K_{electron} = hf - hf' = h(f - f')$$

(b) First find the speed of the electron:
$$K = \tfrac{1}{2}mv^2 \rightarrow v = \sqrt{\frac{2K}{m_e}} = \sqrt{\frac{2h(f - f')}{m_e}}$$

Next solve for the momentum:
$$p = mv = m_e\sqrt{\frac{2h(f - f')}{m_e}} = \sqrt{2m_e h(f - f')}$$

(c) $\lambda = \dfrac{h}{p} = \dfrac{h}{\sqrt{2m_e h(f - f')}} = \sqrt{\dfrac{h}{2m_e(f - f')}}$

You can find these problems, and more, in worksheet form, ready to print, directly from the APlusPhysics.com site at http://www.aplusphysics.com/ap2/ap2-supp.html

Index

electromagnet 175
electromagnetic force 72
electromagnetic induction 177, 178
electromagnetic spectrum 188, 196–198
electromagnetism 166–186
electromagnets 166
electromotive force 141
electrons 70, 72, 74, 107, 116, 123
electronvolts 97–106, 252
electroscope 77, 78
electrostatic constant 80, 90
elementary charge 70
emf 141
emission spectrum 255
energy 2, 3, 188, 259
  binding 261
  conservation of 125, 129, 260
  electrical 124–126
  kinetic 178, 242, 252
  of a wave 189
energy level diagrams 252–254
energy levels 70
entropy 66
equilibrant 21
equipotential lines 98
equivalent resistance 130, 135

F

Fahrenheit 41
ferromagnetic 160
fields
  electric 82, 97
  magnetic 161–164
first law of thermodynamics 59
fission 261
fluid 24
fluid continuity 35
focal point 211
force
  electrical 90
  electrostatic 80–85, 86
  field 85–95
  fundamental 72, 73
  gravitational 90
  magnetic 160–161
frame of reference 268
Franklin, Benjamin 116
frequency 193–195, 206–208
fundamental forces 72, 73

G

gamma decay 264
general relativity 269
General Relativity 269
Germer, Lester 248
gravitational force 72
gravity 90

grounding 78
ground state 252
guitar 203

H

hadrons 73
half-life 264
heat 45
Heisenberg Uncertainty Principle 258
Hertz 193
Huygens' Principle 229
hydraulics 32
hydrogen 252

I

ideal gas law 53
index of refraction 216–221, 219
induction 78, 177
insulators 75
internal resistance 141
inverse square law 81–85, 90
ions 259
isobaric 61
isochoric 61
Isolines 98
isothermal 61
isotherms 61
isotopes 259

J

joule 96

K

Kelvins 41
kilogram 10–11, 90
Kirchhoff, Gustav 129
Kirchhoff's Current Law 129
Kirchhoff's Voltage Law 129

L

law of conservation of charge 76
law of refraction 219
lens equation 212, 224
Lenz's Law 178
leptons 73, 74
lift 36
light
  speed of 194, 216, 260
lucite 220

M

magnetic domain 160
magnetic induction 178
magnetic permeability 162
magnetic quantum number 257

superposition 200–202, 229

T

temperature 40
terminal voltage 141
Tesla 161
thermal conductivity 47
thermodynamics 58
third law of thermodynamics 67
Thompson, J.J. 250
Thomson, George 248
threshold frequency 242, 243
topographic map 98
Torricelli's Theorem 37
total internal reflection 221
transmission 208
trigonometry 13–15
tritium 262
troughs 189
tuning fork 192

U

ultraviolet catastrophe 241
unit conversions 11–12
universal gas constant 53
universal gravitational constant 90
universal mass unit 261

V

vectors 15–16, 81
  components of 16–20
  equilibrant of 21
velocity selector 167
VIRP table 130–134, 135–141
virtual image 210
voltage vs. current graph 121
voltaic cells 123
voltmeters 126
volts 96, 106

W

watts 125
wave equation 194–195, 216
wave function 258
wavelength 189
  de Broglie 248
wave-particle duality 240–241
waves
  crests 189, 201
  electromagnetic 188, 194, 196–198
  interference 200–202
  longitudinal 188
  mechanical 188
  phase 189, 198, 202
  radio 188, 197
  seismic 188

sound 190
standing 202–205
transverse 188, 190
troughs 189, 201
velocity 194–195
X-rays 188, 197, 248
weak nuclear force 72
wires 119
work 3, 96, 124–126
work function 242

X

X-ray diffraction 233

Y

Young's Double-Slit Experiment 230, 248
Young, Thomas 230

Z

zeroth law of thermodynamics 58

Made in the USA
Lexington, KY
13 July 2018